Cultural Psychology of Recursive Processes

A volume in
Advances in Cultural Psychology
Jaan Valsiner, *Series Editor*

Cultural Psychology of Recursive Processes

edited by
Zachary Beckstead
Grand Valley State University

IAP
INFORMATION AGE PUBLISHING, INC.
Charlotte, NC • www.infoagepub.com

Library of Congress Cataloging-in-Publication Data

A CIP record for this book is available from the Library of Congress
http://www.loc.gov

ISBN: 978-1-68123-018-4 (Paperback)
978-1-68123-019-1 (Hardcover)
978-1-68123-020-7 (ebook)

Printed and bound by CPI Group (UK) Ltd, Croydon, CR0 4YY

CONTENTS

SERIES EDITOR'S PREFACE

FROM RECURSIVITY TO PRECURSIVITY

Making Psychology Generative

Despite keeping up our identities, we are never the same. When taking a moment in our long life course and ruminating about times past, one may feel strange looking at oneself represented by one's baby pictures in a family photo album. "Could I ever have been that baby?" is the curious question that can be asked. The answer is "yes," and its explanation is in the topic of this book—*recursivity*. We notice that we are no longer the same as we look back in time. Yet it does not matter, beyond the moments of melancholy about the youth we have left behind. At any moment in our life course, we live in the opposite direction—we look forward to what we have not yet experienced. We do it on the basis of our past experiences—the past guides our future. Yet we are constantly in the process of constructing the present out of the future. But how is that possible? We cannot return to the future that has not yet become actualized. Once it does, it does not repeat the past.

Recursivity is a strange concept. It implies return to a previous state of affairs—yet in human lives, no such returns are possible. What it is turns out to be a complex dialogue between mathematics, para-mathematics, linguistics, and common sense (see Lee Rudolph's comprehensive coverage in Chapter 2). While closely linked with the philosophical notion of *infinity*,

Cultural Psychology of Recursive Processes, pages vii–x

the treatment of *recursivity* in the formal systems of mathematics seems paradoxically abstracted from time. Time is implied (through infinity) yet not considered in any other substantive way than the equivalent to space in which infinity is constantly in the making. Time is theoretically needed in mathematics only to give a spatialized dimension for infinity. In psychology that is not sufficient.

What would happen if instead of time as a spatial dimension, the emerging theory of recursivity in cultural psychology would consider time as irreversible and intertwined with the recursive (never the same, yet similar—Sovran, 1992) focus towards the future. That would have two implications—the importance of the *processes of imagining* (the possible) futures (see also Chapter 6), and the recognition of absolute uniqueness of each and every instance of the past (see Rudolph, chapter 2, especially the reference to Poincaré's insight into what he called "*raisonnement par récurrence*"). The past is given to us by a sequence of unique life events of determinate kind, while the future is imagined as a field of infinite possible events that may happen under some conditions, yet also do not need to take place. Recursivity is thus a characteristic of moving towards the future. We may be better off calling such version of *contextualized recursivity* by a slightly altered term—*precursivity*—that would keep our investigative minds alert in seeing the processes of moving towards the future in the present. In our book series that focus will be covered by the Trajectory Equifinality Approach (Sato, Mori, & Valsiner, 2015).

Precursivity would fulfill all the four requirements that Zachary Beckstead in his introduction outlines. It is *self-referencing*—through projecting the present onto the field of anticipated future field, part of which may be desired and the rest tolerated, or even avoided. Such future projection builds on imagining, which feeds back into the present to aid in the move towards the future. The self-referencing process is between one's past and the expected future—guiding the transformation. The looping starts from the projected future, feeds back into the present, and leads to the *anticipatory action* towards the future.

The precursive process leads to increased complexity. It brings the focus on *generativity* to the center of psychological thinking. Generatvity replaces the ontological focus that has dominated psychology over its two centuries of existence. While anticipating the future, pre-adaptations emerge that give meaning to what is being anticipated. The depth of the dynamic semiotic hierarchy of signs (Valsiner, 2014) is a way of coping with the task of moving to the yet-unknown future. Our aspirations, values, fears of potential accidents or of other destructive situations, all operate through increasing complexity that culminates in various forms of art as constructive externalizations. In the infinite interiors of our subjectivity, such complexity may take the forms of moral suffering, deep dedications in love or patriotism, and in the search for oneself through recurrent visits to the couch of the psychoanalyst. Both practical psychology and fortune telling (Aphek & Tobin, 1990) thrive on the universality of precursivity.

However, human beings not only construct; they also destroy. The constructive complexity that we can observe in the arts, sciences, and other domains of public life is paralleled by the anticipatory destructive complexity—also part of the precursive process. Humankind excels in the invention of tools for destroying fellow human beings—aside from undesired pests and desired profits from cutting down our forests. Chemical, nuclear, biological, and other kinds of weapons are one of the precursive achievements of humankind that function psychologically even if rarely or never tried out in practice.

Finally, the precursive process is holistic. This follows from the totality of the organismic pre-adaptation to the not-yet-knowable future. Cultural psychology needs to return to the traditions of *Ganzheitspsychoilogie* (Diriwächter & Valsiner, 2008) of a century ago to find ways to make sense of this holistic nature of human dealing with the *Umwelt.* Qualitative mathematics (Rudolph, 2013) may provide some useful leads here. Psychology at large is undergoing a qualitative turn, and a new science of holistic abstract kind is slowly emerging.

Recursivity is not only an abstract concept; editing a book is based on the experiences of the past, yet it precursively creates the future. This book itself is an example of recursivity—the idea for it was born in the activities of our "Kitchen Seminar," a weekly meeting of discussion of projects-in-progress. That seminar—originated in 1997 at Clark University and transferred to Aalborg University in 2013 as the core of the new Niels Bohr Centre of Cultural Psychology—has led to spontaneous emergence of a number of scholarly publications. *Relating with environments*—edited by Rose Sokol Chang (2009) —was the first of these, and the present one on recursivity the second. The idea for putting together a scholarly book on the topic emerged in one of the videoconferences of the "K-seminars" in 2010 with Adolfo Perinat from Barcelona who at that time joined us from Cali—the recursive move of argument from North America to Europe, and back, set the process moving. Other contributors joined as the book idea was progressing, while others dropped out. The role of Mariela Orozco from Universidad del Valle in Cali, Colombia in the beginning phase of editing the book and her enthusiasm about it needs to be most positively acknowledged. It is more than unfortunate that the condition of her health did not allow her to complete her editorial role.

Cultural psychology has been an arena for new intellectual adventures in psychology over the past two decades. The present volume extends this search for novelty in an otherwise unimaginative discipline (Valsiner, 2012) to the realm of (p)recursive processes. Opening our minds to new ways of thinking is a difficult task in any society, so the readers of this volume are up to a delightful intellectual challenge.

—Jaan Valsiner
Aalborg, Denmark

REFERENCES

Aphek, E., & Tobin, Y. (1990). *The language of fortune telling.* Amsterdam, The Netherlands: John Benjamins.

Chang, R. S. (Ed.). (2009). *Relating to environments.* Charlotte, NC: Information Age.

Diriwächter, R., & Valsiner, J. (Eds.). (2008). *Striving for the whole: Creating theoretical syntheses.* New Brunswick, NJ: Transaction Publishers.

Rudolph, L. (Ed.). (2013). *Qualitative mathematics for the social sciences.* London, UK: Routledge.

Sato, T., Mori, N., & Valsiner, J. (Eds.). (forthcoming). *The trajectory equifinality approach.* Charlotte, NC: Information Age.

Sovran, T. (1992). Between similarity and sameness. *Journal of Pragmatics, 18,* 329–344.

Valsiner, J. (2012). *The guided science: History of psychology in the mirror of its making.* New Brunswick, NJ: Transaction Publishers.

Valsiner, J. (2014). *Invitation to cultural psychology.* London, UK: Sage.

INTRODUCTION

ENTERING THE RECURSIVE REALM

Zachary Beckstead
Grand Valley State University

Human living is a magnificent and complex process. While we are situated and enmeshed in innumerable environments and contexts, we attempt to remember what happened minutes, days, or years ago in an effort to enjoy our previous experience, refute an accusation, or make sense of our current life circumstances. We also spend time endlessly worrying or richly anticipating what will happen in our immediate or distant future (and our death). We play, fantasize, fight, argue, strategize, help, and engage in thousands of actions in a day while also admiring a beautiful sunset or being disgusted by the sights and smells that surround us. Furthermore, these actions and relationships are largely, though not exclusively (see Guimaraes & Cravo, Chapter 5, this volume), mediated through the use of language, which itself is a very interesting and complex system. One unifying claim among the diverse chapters in this book is that recursion and recursive processes are at the core of these complex social and psychological processes. Recursion is bound up with the notion of *returning, reexamining, reflecting,* and circling back, and these processes allow for human beings to simultaneously distance themselves from the here-and-now settings while being im-

Cultural Psychology of Recursive Processes, pages xi–xxxi

mersed in them. Thus, the purpose of this book is not simply to celebrate the complexity of human living, but to extend the notion of recursion, recursivity, and recursive processes into the realm of social and psychological processes beyond the arenas in which these ideas have currently thrived.

The concepts of recursion and recursive processes can easily be found in the fields of linguistics and mathematics but are quite rare in other areas of the social and human sciences (see Corballis [2011] for an exception). Rudolph (this volume) provides an excellent historical overview of recursion in mathematics, but he finds that there is no concensus on what it means and entails in the social sciences. The concept of recursion is multifaceted but also defined in very different ways such that there seem to be as many definitions as there articles on recursion. Furthermore, as Long, Yarrison and Rowland note in their chapter on *imagined recursivity*, few articles can be found in psychological journals in the social sciences even if many of the ideas and approaches found therein are inherently recursive. Similarly, discussions of recursion are almost completely absent in introductory textbooks in social psychology, sociology, and psychology. Thus, while recursion has been claimed to be a key property of the human psyche that differentiates us from our mammalian relatives, it has often been underdeveloped and implicit in fields like psychology, sociology, and related disciplines.

The aim of this introduction, however, is not to provide a historical review of recursion, recursivity, and recursive processe; rather, it is to provide the reader with an overview of some of the meanings and history of this rich and underutilized (at least within the social sciences) concept. First, I will briefly provide some more informal illustrations and examples of recursion and recursive processes in everyday life and the media. Second, I offer a brief review of the concept of recursion and recursive processes in different scientific and humanistic fields before turning to more formal definitions.

WHAT IS RECURSION?

Informal Illustrations of Recursion

> Recursion (ri-kur'zhen) *noun.* If you still don't get it, see *recursion.*

The above "dictionary definition" is one take on recursion that can be found in numerous books and even the illustrious Wikipedia. For those who take this definition seriously, following these instructions could easily lead to an unending quest.

Another common example of recursion occurs when two mirrors face each other. This positioning creates an effect where the images are embedded within each other and are repeated infinitely. While each nested image appears to be the same, there are differences in these images. For instance, each succeeding image is smaller than the proceeding image. Importantly, here we see a repetition of the similar and not the same form.

We can find recursive images in works of art and advertisements. The Droste effect occurs when, in a picture or painting, a picture appears within itself over and over again. This was made famous by the Dutch cocoa company, Droste, and their advertisements of a nurse holding an object, a serving tray with cocoa and a mug that has the same image. Like a series of Russian Dolls, the images are nested within each other.

Literature has also contributed to the recursive well. Author and mathematician Augustus de Morgan wrote (as cited in Corballis, 2011, p. 3) about fleas in a delightfully frightening way:

> Great fleas have little fleas upon their backs to bite 'em,
> And little fleas have smaller fleas, and so ad infinitum.
> And the great fleas themselves, in turn, have greater fleas to go on,
> While these again have greater still, and greater still, and so on.

Movies also provide some humorous illustrations of recursion. For instance, in the movie *The Princess Bride* (Lear & Reiner, 1987), there is "duel of wits" where the Man in Black presents the kidnapper of the princess, Vizinni, with two glasses of wine and puts poison in one. Vizzini is then given the task to choose which has poison and which does not and then drink from that cup (while the Man in Black will drink from the other cup). Here ensues the decision process of the kidnapper:

Man in Black: All right: where is the poison? The battle of wits has begun. It ends when you decide and we both drink and see who is right and who is dead.

Vizzini: But it's so simple. All I have to do is divine from what I know of you: Are you the sort of man who would put the poison into his own cup or his enemy's? Now, a clever man would put the poison into his own goblet because he would know that only a great fool would reach for what he is given. I am not a great fool, so I can clearly not choose the wine in front of you. But you must have known that I am not a great fool, so I can clearly not choose the wine in front of me.

Man in Black: You've made your decision then?

Vizzini: Not remotely! Because Iocane comes from Australia. As everyone knows, Australia is entirely peopled with criminals. And criminals are used to having people not trust them, as you are not trusted by me. So I can clearly not choose the wine in front of you.

Man in Black: Truly, you have a dizzying intellect.

Vizzini: Wait 'til I get going! Where was I?

Man in Black: Australia.

Vizzini: And you must have suspected that I would have known the poison's origin, so I can clearly not choose the wine in front of me!

Man in Black: You're just stalling now.

Vizzini: You'd like to think that, wouldn't you? You've beaten my giant, which means that you're exceptionally strong . . . so you could have put the poison in your own cup, depending on your strength to save you, so clearly I cannot choose the wine in front of you. But you've also bested my Spaniard, which means you must have studied, and in studying, you must have learned that man is mortal, so you would have put the poison as far from yourself as possible, so I can clearly not choose the wine in front of me.

Man in Black: You're trying to trick me into giving away something. It won't work.

Vizzini: It has worked! You've given everything away! I know where the poison is!

Things did not end well for Vizzini. Unfortunately for him, the Man in Black had poisoned both cups of wine and was able to survive because he had built up a tolerance to the poison prior to the duel. Vizzini, on the other hand, immediately died and ended his recursive monologue.

Each of these examples highlights different characteristics of recursion, yet each one conveys a meta-level quality that is intrinsic to the idea of recursion. This can be articulated in the statement that through recursion one can think about thinking, and one can think about thinking about thinking, and so on. These examples also illustrate how recursive looping that can be "extended indefinitely to create sequences or structures of unbounded length or complexity" (Corballis, 2011, p. 6). What is important to note, for the purposes of this book, is that recursion allows for depth and complexity in human thought and allows us to transcend the here-and-now context in order to return to our past and plan future actions. The recursive nature of human thought allows for players in chess to think of and anticipate a wide array of moves from their opponent and countermoves in response (Corballis, 2011). Thus, recursion and recursive properties allow for flexibility and coordination in human action. Recursion is thus an expansive concept that cuts to the core of what it means to be not only the thinking animal, but also a social animal.

Note: While in theory these thoughts or images could go on *ad infinitum*, the capacity of the human mind is limited and therefore infinite regress does not come into play.

SPECIFYING AND ELABORATING RECURSION

Recursion is a very abstract concept with both literal/common sense and more technical/academic meanings. The older and more "literal" definition of recursion implied a return or backward movement of phenomena such as the tide or blood, or the gaze of the individual upon an object (i.e., returning to look at x once more). Perinat (this volume) traces the etymology of recursion to the Latin verb of *recurrere* (to return go back or return). The implication is that recursivity occurs when something retraces its prior movement and returns to the beginning. There is a loop (i.e., self-reference) that occurs in this movement. Reiteration, as Perinat also notes, has a similar relationship to recursion but implies a discursive return. As Rudolph (Chapter 2, this volume) notes, the more elaborate and technical meanings of recursion come from mathematics and linguistics. Below I will briefly chart out how recursion and recursivity have been defined and explored in the fields of mathematics, linguistics, hermeneutics, and more recently in psychology and other fields within the human and social sciences.

Mathametics and Cybernetics

Recursivity in mathematics has a long and storied history and, as Rudolph demonstrates in this volume, is more formally defined than it is in the other human and social sciences. In terms of mathematics, then, recursivity exists when phenomena can be defined by two properties: (1) a simple base case and (2) a set of rules that reduce all other cases toward the base case. Corballis (2011) and others have used the Fibonacci series as an example. It is comprised of two base cases and then a recursive definition:

fibonacci(0) = 1 [base case]
fibonacci(1) = 1 [base case]
fibonacci(n) = Fibonacci($n - 1$) + Fibonacci ($n - 2$)
[where $n > 1$]. [recursive definition]

Interestingly, Fibonnaci, who was an Italian mathematician, utilized this to project the proliferation of a population of rabbits (Corballis, 2011). What makes this equation recursive is that, in addition to the base cases, each number (n) is the result of the two previous numbers.

Corballis gives another illustrative example of how recursion functions in mathematics. Drawing on the notion of *factorials* ($3! = 3 * 2 * 1$ equals 6), he shows how each set can be depicted by using two defining equations:

$$0! = 1$$
$$n! = n^* \ (n-1)! \ [\text{where } n > 0]$$

As he notes, "the second equation is recursive in that a factorial is defined in terms of a factorial" (Corballis, 2011, p. 5). The notion of self-reference constitutes recursion, or in other words, "the right hand side contains, in some way, the entity A that is present in the left hand-side" (Sawada & Caley, 1993, p. 3). Like mirrors facing each other, the right and left side reflect and are embedded in each other.

Sawada and Caley point out how a simple mathematical function can be depicted recursively. They offer the mathematical function that gives the sum of the first n odd integers as an example:

$$\mathrm{S(n)} = 1 + 3 + 5 + 7 + \ldots + 2$$

In order to emphasize this equation recursively, they change the equation in the following form:

$$\mathrm{S}(n) = \mathrm{S}(n-1)$$

Scholars have also defined recursion in abstract mathematical terms where recursion is closely linked to "mathematical iteration." Here recursion is exemplified when the output is fed back into the input, and this is done over and over again and "each new solution (y) becoming input (x) for the next repetition of the equation" (Clifford, Friesen, & Jardine, 2001, p. 3). Iterative procedures can be found in computational mathematics and cybernetic feedback systems that maintain homeostasis in organisms. Some scholars differentiate iterative from recursive processes; however, the key question is whether or not iterative processes lead to added complexity. For instance, the humorous definition at the beginning of the chapter is simply repeated over again and leads to no added complexity (Corballis, 2011). In contrast, the physical act of doing push-ups may be repeated, but each successive act creates a new experience for the one performing the push-up (see Watzlawik et al., Chapter 6, this volume).

Additionally, in mathematics, something is considered "recursive" if it is defined (or built up) in cycles, starting from a "base case" and allowing the definition to have a circular appearance because it *appears* to refer to itself. However, this not really so because in any specific case, after enough unwinding, you get back to the base case, which is ideally the "simplest" thing of the kind being defined.

Recursion and Linguistics

Many linguists have considered recursion to be at the core of language (Chomsky, 1957). The importance of language and the "concern that led to recursion" can be traced back to Descartes (1637/2003) and Wilhelm von Humboldt. Descartes suggested that only a human being has the ability to "arrange its speech in various ways in order to reply appropriately to everything that may be said in its presence, even as the lowest type of man can do" (Descartes, 1637/2003, p. 38). Wilhem von Humboldt echoed this notion by stating that only humans have the capacity to "make infinite employment of finite means" in language (1836/1999, p. 91). Breaking with the behaviorist model of language, Noam Chomsky famously argued that human thought and language originated from operations applied recursively. Simple units of words can be merged and embedded with other units to form larger entities, and this process can be extended in increasing complexity and hierarchical organization. For instance, noun phrases can be created from other noun phrases as they are recursively embedded: "One can create new noun phrases by placing the word *beside* between any pair: *the dog beside the tree, the cat beside the lake*" (Chomsky, 1957, p. 6).

Furthermore, linguists in the tradition of Saussure (1916/2011) point out that words and utterances derive their meaning from their relationship to other words and groups of words. Recursivity, in this sense, is related to the interrelationship of symbols and the meta-reflexive ability of the human mind to discover and grasp meaning and higher-order relationships and interconnections between parts (see Perinat, 2007). Human beings, then, can not only "decode" language, but reflect upon it.

Psychology and Recursion

Recursion, recursivity, and similar ideas typically enter into the field of psychology through the field of linguistics. As previously mentioned, it is recursion that is seen as the distinguishing faculty of human beings. Recently, the question of whether all human languages are fundamentally recursive has been a hot topic in psychology and anthropology (see Everett, 2005 for a discussion on Piraha language). This controversy emerged when a young missionary turned atheist visited a small Brazilian tribe and discovered that they live in the present moment and do not reference the past or future. Hence, they have no creation myths or sense of history. They are also lacking in words for colors and numbers. The Piraha express their language nonrecursively, or through nonembedded clauses and phrases. Examples of nonrecursive languages notwithstanding, most languages do feature recursive processes and aspects and have garnered most of the attention of psychologists. Indeed, language disorders such as aphasia can be understood through the lens of recursion (Bánréti, 2010).

Michael Corballis (2011) has extended the idea of recursivity and linked to thinking, memory and "theory of mind." Corballis argues that not only is language recursive, but so is thinking. Indeed, he makes the case that recursion exists in language to express recursive thinking, or nonlinguistic thought. He also points out how memory is intertwined with recursion, as human beings constantly "insert" or "embed" the past into the present and imagine future realities through mental time travel. In other words, recursive processes are implicated as we relive the past and anticipate the future. For instance, we relive a vacation taken long ago or start working out at the gym in spring in order to be prepared for summer. This suggests the very functional aspects of recursion in relation to human adaptability. Moreover, he also argues that recursivity is the key to understanding the intentions and beliefs of others. This ability to "read the minds" of others is referred to as "theory of mind" in psychology and has implications for our lived experience and disorders such as autism (Corballis, 2011; see Chapters 7 and 8, this volume).

Recursion and the Humanities—Hermeneutics

Recursion and recursive processes are also implicitly related to hermeneutics—the theory of interpretation—where notions of *returning* and *re-examining* play a central role. Clifford et al. (2001) argue that while most understandings of recursion anchor it as a process of the self-looping back on itself to create self-awareness, a hermeneutic understanding of recursion sees it not as an act of mind the self-looping back on itself, but it is constituted by the very ontological nature of being or the conditions of human existence. Ontological hermeneutics posits that the world is dynamic and constantly changing, that "things" are their multiplicity, and a part can only be understand in relation to the whole. The implication is that things (concrete desks and chairs and less tangible things such as words) are not static entities that we take different perspectives on, but the fundamental (hence ontological) aspect of the thing is its variety and possibility. The hermeneutic circle emphasizes tracking back and forth between the whole and the part (or moving between self-understanding and understanding of the world) and how our *prejudices* predispose us to encounter and make sense of the world in a certain way. This is not an activity conducted only by academics, but rather it is a fundamental aspect of human existence that we confront. Hence, understanding is a mode of being and is never final. This holistic emphasis is also found in the writings of Gregory Bateson and his discussions of recursive epistemology.

TOWARDS A WORKING DEFINITION?

As noted above and in Rudolph's chapter in this book, there are many definitions of recursion, and these can differ greatly. Additionally, Sawada and Caley (1993) raise the question of whether recursivity is inherent within the phenomena or if it is up to the observer to "see" and model phenomena as recursive or nonrecursive. Discussing recursivity in relation to Bohr's notion of complementarity, Sawada and Caley state:

> We presume that there are no particular entities or phenomena which in themselves are recursive; rather, it is the way that we as observers (mathematicians, anthropologists, philosophers, programmers, poets, narrators, etc.) depict these entities through our descriptions that makes them recursive. It is not the phenomenon itself, but our choice in the way we depict the phenomenon that is our critical in establishing recursion. (pp. 2–3)

The multiple meanings of recursion, and their relationship to similar notions of repletion, iteration, reflection, induction, transitivity, and branching, in addition to ontological questions of recursion (Sawada & Caley, 1993, p. 3), are certainly a challenge for our volume since we are operating under the stated goal that is to clarify and extend the notion of recursion into the realm of social and psychological processes. However, instead of forcing authors into adopting a singular definition of recursion, we have asked each author to specify what definition he or she is working with and elaborate on the conceptualization of recursion.

WHAT RECURSION IS NOT

Deciphering what recursion is, and thus how it can be fruitfully applied to psychological phenomena, can be done, in part, by examining what recursion is not. Corballis (2011) notes that recursion is often confused with *repetition* and *iteration.* An example of repetition is can be found in the numerous lists we create. We can create a grocery list and add item after item to it. Similarly, we can also repeat a mantra over and over again. While repetition can create an infinite sequence of utterances or thoughts, simply adding new information to them does not "embellish or qualify" these sequences. The sentence is changed by adding new nouns and phrases, but it is not related to or driven by the previous elements of the utterance or thought and is therefore not recursive. It conveys information but does not necessarily lead to transformation of the sentence. Yet, as Watzlawik, Schachter and Cunha (this volume) point out, what often seems to be *mere repetition* often includes novelty in the process. Likewise, *iteration* involves processes that are repeated, but they do not involve added complexity. Thus, the "dictionary" definition of recursion included at the beginning of this chapter is iterative and not recursive in that nothing new is added and one keeps going around in a loop without transforming the structure in the process.

While recursion is not simply *repetition,* Watzlawik et al. (Chapter 6) al Guimaraes and Cravo (Chapter 5) persuasively demonstrate that when it comes to human beings there is no such thing as *mere repetition.* Repeating an action, such as a push-up or sounding the same note on an instrument, may involve behavioral repetition, but *for the person* the action is never the same. Watzlawik et al. point out that even though the same behavior may reoccur, the way the person creates meaning changes. They offer the example of push-ups to illustrate that as they are repeated the person may be "fresh and motivated at the first set, but [may have] to increase his or her effort with every push-up" (p. 162). While the same or similar action is repeated, it gains added significance from the past (i.e., number of previous push-ups or previous experiences) and future, anticipated events or goals (i.e., to impress a significant other with a new physique). Thus, repeated actions and utterances are embellished, modified, and changed from the perspective of active, meaning-making agents.

CHARACTERISTICS OF RECURSION

Pinning down and constructing a definition of recursion that is widely accepted is virtually impossible; nevertheless, we can at least highlight some of the shared features of these various definitions. These features, elements, or characteristics of recursion are themselves interdependent and include the following: self-reference, transformation, involving whole < > part relationships. Below is a brief elaboration of these characteristics.

Self-Referencing

Almost all notions of recursion involve the concept of self-referencing, level-stepping, and circularity. Citing Vitale (1989), Sawada and Caley (1993) posit that any definition of recursion requires that "there is an element of permanence (circularity) in it: the right hand side [of the equation] contains, in some way, the entity A that is present in the left side"

(p. 3). Of course this requirement is related to the mathematical sense of recursion; however, formal and informal definitions are characterized by this sense of looping and return. The English language is self-referential, as can be illustrated by terms such as "this" or "that"—used in a sentence they refer to themselves.

Transformation

While self-referencing is crucial to any notion of recursion, the importance of this looping is that it creates change. Thus returning (in the literal and metaphorical senses), reflection, and circling back do not simply involve an act being repeated in whole. Instead, novelty and transformation are inherent in any recursive process. Processes are repeated in self-similar ways, but never in the same ways.

Increased Complexity

Change and transformation that result as a part of recursive processes thus lead to increased complexity. Returning to the phenomenon of placing two mirrors together, we see that this action creates a seemingly endless and complex series of images that is suggestive of infinity. I recall being endlessly fascinated by looking into these mirrors when I was a child getting my hair cut at the barber. In a more sacred and ritual context, two mirrors are placed toward each other in the temple room where members of the Church of Latter-Day Saints marry for "time and all eternity." Bride and groom are placed before an altar, in between these mirrors facing each other, and are instructed to look into them and into eternity. Rudolph notes the horror that notions of infinity have caused some mathematicians—sterile infinite regress—but this very human notion seems to be related to our recursive capabilities.

Holism

Additionally, almost all notions of recursion and recursive processes explicitly or implicitly involve part < > whole relationships that are constantly changing. As mentioned above, this is clear in the context of language where, according to Saussure (1916/2011), the meaning of words derived from their position and relationship to other words. These part < > whole relationships can be found between the person < > environment, past < > present < > future, and so on.

QUESTIONS TO BE EXPLORED (GOALS OF THE BOOK)

Self-referencing (looping), transformation, complexity, and holism constitute the core facets of recursivity and provide the basis for which authors in this book explore and elaborate this concept. Each author takes a particular phenomenon and tries to illuminate this through the lens of recursion and recursive processes. Thus recursion provides an important conceptual framework for extending our understanding of a vast array of social and psychological processes. We focus on the systemic nature and transformative aspects of interrelated parts and process nature.

THIS VOLUME

The genesis of this book can be traced back to a quite innocent discussion in a kitchen at Clark University about recursive properties of psychological phenomena. Details of this conversation are fuzzy, yet what emerged was the notion that recursivity was itself a fuzzy and complex idea worth pursuing. Developing this volume on recursivity has posed its own unique challenges and for us has been truly a voyage into personally unknown and uncharted territories (for me as well as for many of the contributors to this book). We felt confident in saying that phenomenon X or Y was recursive, but when pinned down on what exactly recursion was, its characteristics, and what looking at phenomena from this perspective offered, we struggled for clear and satisfying answers. Recursion seemed to be referenced often in regards to mathematics and linguistics and only occasionally in the social sciences. Yet as we searched the academic literature, the references we discovered only contributed to our sense of increasing ignorance. Similar to being able to see more stars in the Milky Way by looking through peripheral vision, our direct gaze at recursion seemed to only problematize our understanding and in a sense made us blind. In spite of our considerable limitations, we pursued this topic convinced that collaboration with fellow adventurers would yield both theoretical and empirical innovations. We thus saw pursuing and developing the notion of recursion in the human sciences as an opportunity to look to concepts from other disciplines for stimulation, but also to go beyond simply transporting ready-made understandings from the natural sciences.

OVERVIEW OF CHAPTERS

In the first chapter, Adolfo Perinat offers a rich and poetic review and extension of recursivity and recursive processes that moves from paradoxes and parentheses to observation, play, ritual, and theatre. He argues that these phenomena, typically reserved for anthropologists and ethnologists, are at their core recursive processes or, better stated, they rely on the recursive capacity of the human mind. Perinat notes that the etymological roots of the term recursivity can be traced back to the Latin verb *recurrere* (to go back and return) meaning "to retrace the same path," which entails to and fro movements as one returns to the beginning. According to Perinat:

> Recursivity or, if you prefer, recursive processes are manifestations of human cognitive activity characterized, in a first rough approximation, by the fact that the mind's discourse executes a back and forth movement or makes "jumps" between the different domains in which it operates.

Recursive processes are therefore are evident in paradoxical statements that are ubiquitous in human history. Utterances such as *the essential lies in what is not essential, never say, "never,"* and *we must not tolerate intolerance* are statements that loop back, retracing their path back to the beginning, and seemingly contradicting themselves. These statements violate the principle of noncontradiction; however, human beings have been using and making sense of such statements for years, to the horror of logicians. Additionally, Perinat notes that parenthetical statements and footnotes are also inherently recursive. Both parentheses and footnotes interrupt the flow of discourse and then allow for a return while providing for a "productive expansion of the train of thought" (p. 2). Here we begin to see the main points of recursivity that Perinat develops: "that the recursive mind crosses borders; it is a "smuggler" that moves with ease between territories" (p. 2).

These ideas are fleshed out with the notion of the *observer*. According to Perinat, the observer does not simply stand outside of an event and record it; rather, the observer is *inside* and *outside* of the event and is an *actor* and as well as *observer*. They are able to shuttle back and forth between both perspectives and mental domains. This back-and-forth movement, the creation and eradication of boundaries and distinctions, giving semantic meaning to actions, are each recursive activities that allow for the emergence of rich human activities such as play, ritual, and theatre.

The chapter can be seen as a mediation on recursivity that elaborates on recursion through successive approximations starting with paradoxes, parentheses and footnotes, moving into the activities of observation, framing (a la Bateson) and play, then turning to the fascinating cultural activities of ritual and theatre while slowly and insightfully drawing out the self-referential, reflective and transformative aspects of recursivity. Again, the to and fro activity of constructing and crossing domains and frames are central to recursive processes. Finally, Perinat raises the thorny question of reality: whether the particular aspects of the reality defined as play, theatre, ritual, myths and so on, are part of our life as legitimately as the reality regarded in a vulgar sense as "real."

Following Perinat, Lee Rudolph (in Chapter 2) offers a thick investigation into how recursivity, recursion, and similar concepts are understood and employed in mathematics and the human sciences. In mathematics, notions of recursivity can be traced back to base cases and lead to ideas of aposiopesis, infinity, and well-foundedness. With a very thorough lens (Rudolph modestly describes his exploration as a "survey"), he explores both the "horror of infinity" engendered by recursion and its twin, an enthusiasm for infinity. For mathematics, the history and use of recursivity, recursion, and similar ideas is clear, if also complex. However, when Rudolph peers into the meaning and use of recursion and recursivity in the human and social sciences, he finds "a tower of Babel, a confusing braid, a handful of roots" (p. 44). He notes and provides 25 different definitions of recursivity, some of which share similarities, but many of which offer contradictory meanings. As with recursivity in mathematics, Rudolph traces their use to a few base cases. He deftly demonstrates how the meaning of recursion and recursivity in the human sciences can be linked back to cybernetics, systems theory, mathematics, and linguistics. Rudolph's exploration highlights a key challenge of this volume: How are we to understand and utilize such a complex concept? Do we need to have a clear and agreed-upon definition of recursion to avoid this tower of Babel? As mentioned above, we have discouraged authors from using a single definition of recursivity or recursion while encouraging clarity of how they are defining and using these notions. Still, Rudolph's chapter is, on the one hand, a necessary reminder of the complexity of recursivity and importance of clarity of definitions. More importantly, this chapter excels in presenting a historical overview of these terms and also analyzing their implications. Rudolph's "survey" of recursivity and recursion is extremely comprehensive and scholars in many disciplines and especially those with an interdisciplinary spirit like Rudolph should find this chapter challenging and particularly rewarding.

In Chapter 3, Eric Charles explores the role of recursivity plays in the theoretical approaches of radical empiricism and radical behaviorism. He starts with a very simple principle—"things are what you experience when you experience those things"—and, following James, Dewey, and Holt, reapplies this principle over and over again to the output of previous results in order to come up with new and counterintuitive results. Indeed, this very basic principle, Charles argues, serves as a "theoretical acid" that dissolves problematic distinctions (i.e., between subject and object) that have bedeviled modern psychology. Charles both seeks to call attention to James' often neglected fidelity to experience, most notably pursued in *The Variety of Religious Experiences*, and to elaborate on these insights. As Charles notes, James' phenomenology stands in direct contrast to both idealist and dualist epistemologies. Instead of claiming a sharp distinction between subject/object, truth/fiction, physical/mental, James relentlessly and recursively asks, *"What are people experiencing?"* instead of *"What is real?"* Charles first illustrates the potential this approach has for psychology by looking at how a radical empiricist approach deals with simple objects. He draws on Dewey's example where he has different people—a zoologist, a horse-dealer, and others—describing the same horse. While each may feel that he offers a better (i.e., more correct) description of a horse, Dewey denies that there is a more or less "real" account and what we need to focus on are the kinds of experiences are offered. The differing accounts of the zoologist and horse-trader are based on the previous experiences and roles of each. Again, one is not more real than the other. Likewise, Charles provides the example of a bowl of spaghetti and illustrates that *what it is* depends on how it is experienced by the person. Even if we later find out that a bowl of spaghetti is really a bowl of vermicelli, this does not negate our original experience, though it might transform our future experiences. So, in order to address what something is, we need to go back to the radical empiricist principle that interrogates how something is experienced.

Charles proceeds to develop this principle and highlight the recursive aspects of reapplying this principle over and over again in relationship to less concrete objects. In order to do so, he creates an imaginative dialogue between James (radical empiricist), Piaget (rationalist), Descartes (dualist), and Berkely (idealist). The dialogue begins with Piaget's claims that a child shown a new type of chair, with no legs, has accommodated this "chair" into his previous schema of chair. James subjects various explanations and claims by Piaget, Descartes, and Berkely to the question of *what the child experiences.* Each account or category of what the child does or accomplishes offered by the other notable scientists and philosophers *is post-hoc* and part of *their* experience, not the child's (i.e., psychologist fallacy). By returning claims and questions to questions of experience, "recursion acts as a theoretical acid, dissolving down all erected distinctions, such that they must be first understood in terms of immediate experience" (p. 11). Instead of arguing against constructing theoretical frameworks, Charles argues that these frameworks must be built up from and return the realm of experience. Experience is the ultimate authority for the "validity" of any theory.

Finally, Charles elaborates the core principle of "things are what you experience when you experience those things" into very seemingly very abstract realms. Charles notes that James re-applies the same methodological principle of grounding analysis in experience to relationships. Relationships do exist in the ether, but rather are experienced as relationships. James also claims that knowing is a relationship between the *knower* and *knowing.* This insight becomes fodder for radical behaviorism and Holt's new realism. Charles combines Dewey's and James' assertions into a radical behaviorist principle that focuses on what is seen and observed. Taking *mind* as an "observable relation between observable things" then, Charles posits, the radical behaviorist asserts that "When we see another person's mind, we are seeing some pattern of behaviors relative to circumstances. And if that is what we see when we see someone else's mind, then that is what the mind is. Period."

Next, Brooke Long, Fritz Yarrison, and Nicholas Rowland (Chapter 4) explore the recursive role of imagination in stigma management and self-disclosure in the growing group of "nones" (i.e., individuals who do not self-identify with any religion). Long et al. begin their chapter by illustrating the relevance of recursion to the social sciences, in particular social psychology. The authors adopt a very specific definition of recursion, drawn from the humanities, that sees recursion as a process in which mind "loops" back on itself continuously. This notion of recursion is closely linked with Cooley's still relevant and utilized idea of the "looking-glass-self" which, according to Cooley, has three basic principles: "the imagination of our appearance, the imagination of his judgment of that appearance, and some sort of self-feeling, such as pride or mortification." As Long et al. note, this concept is recursive since the mind (or *self* in their words) "loops" back or takes itself as an object of reflection. However, this recursive process entails the self reflecting on the self as others (real or imaged) see and judge the self. This process "takes on a reality or gravity" (p. 124) in social interaction, or even prior to social interaction as the authors emphasize, as individuals tailor their actions based on how they imagine they appear to others or how others will react to them. From these insights, the authors offer their concept of "imagined recursivity" to highlight "the deeply imagined nature of a social psychological definition of recursion." Next, Long et al. examine how "imagined recursivity" (via the "looking-glass-self"), religion, and identity are intertwined. Identity, for the authors, is a "perpetually emergent construction" that involves personal and social dimensions. Religion for many in the United States, in particular, is an essential element of the self that, among other things, tends to bind individuals together. However, for those who do not claim a religious identity and affiliation ("Nones"), this rejection can be potentially stigmatizing. As Americans have become more tolerant of others from various religious backgrounds, atheists and agnostics are still viewed negatively compared to their religious counterparts. Thus, symbolic boundaries exist between religious and nonreligious groups. Based on 35 qualitative interviews, the authors examined how nonbelievers ("Nones") manage the stigma of their "discredited" identities (Goffman, 1963, p. 4). While many cultural resources and strategies exist for neutralizing stigmatized identities, the authors note that these strategies focus on what is happening during social interaction and "fail to harness pre-interaction techniques designed to avoid such interactions in the first place" (p. 127). Long et al. find that when participants are asked whether in a conversation *if* they are asked if they are religious that they would reveal their ("discreditable") nonreligious identities, they respond that they would. However, since this actually never occurs, the authors posit the participants are "hiding without hiding" in that they can avoid revealing their identities but not feel as if they are hiding. This reluctance and stigma neutralization strategy is based on the imagined anticipation of how others would see and judge their revelation. Long et al. close with an exploration of the problems that emerge as individuals engage in the act of "hiding without hiding."

Danilo Guimaraes and Andre Cravo (in Chapter 5) offer an ambitious and integrative approach to understanding how people adapt their environment and understand the "minds" of others through imitation. As they note:

> Imitation can be thought as a recursive process of reproducing someone else's or one's own movement. This process can happen in different levels of awareness. Imitating the other in the environment is one possibility of reacting to it that can be evaluated from a developmental perspective, as Baldwin proposed. Imitation is closely connected with perception and imagination once that the personal imitative action inherently carries some supposition about the movement of the other. (p. 155)

Their insights foreground the biological and subpersonal levels of human existence and their recursive relationship between biological, social, and cultural worlds in the ontogenesis of the person. The framework they develop builds upon the work of James Mark Baldwin, contemporary semiotic cultural psychology, the implications of mirror neurons, and Susan Hurley's shared circuits model. These different approaches are brought into dialogue with each other and allow the authors to examine the biological mechanisms and recursive layers that allow human beings to both distance themselves from the here-and-now settings, and therefore gain control over the environment, and also understand the minds of others. The authors focus on the processes of imitation through the work of Baldwin, Boesch, and Valsiner and through the discovery of mirror neurons. According to Guimaraes and Cravo, we share a world with others through our biological and embodied state and experience via mirror neurons and shared circuits. This similarity (though not sameness or complete unity) underlies our ability to understand the intentions and thoughts of others. That is, when we observe people acting, we are able grasp the intentions behind the actions without reproducing these actions. This approach provides an alternative explanation for what has been referred to as the "theory of mind" and emphasizes the way the physiological or pre-personal level of person < > environment relationship is organized.

Recursivity is highlighted and elegantly described by the authors:

> In such a way, our bodies seem to be inherently addressed to intersubjective experiences. We are intense imitators, especially when young, but still remarkably effected by the action of others as adults. The pre-personal layers discussed by Hurley suggest that an increasing capacity to understand and simulate the action of others can happen even without the use encoding through signs. Because of constant exposure to an environment that is culturally organized by particular persons, the body is biochemically transformed; some neuronal patterns are intensified as some others expire. (p. 157)

This chapter offers both a cogent alternative understanding about how we understand the intentions, thoughts, and beliefs of others and a very sophisticated integration of seemingly divergent theoretical approaches.

Meike Watzlawik, Elli Schachter and Carla Cunha (Chapter 6) interrogate the relationship between identity development and recursivity. Indeed, their exploration of identity exploration emphasizes the crucial, but under-examined role that "looking-back" at the past plays in this process. The authors begin their chapter by clarifying and expounding on their conceptualization of recursivity. Similar to Long et al. (Chapter 4), the authors see recursivity as not simply a repetition of the same process because, for human beings, this is impossible. Citing Driesch, they note that while we might repeat a T.V. program, push-up, or experiment, the *meaning* of "repetitive" act for the person changes since the meaning or content "*carries in itself two accents: one of before and another of already known, which it did not carry when it was possessed first*" (Driesch, 1925 cited in Watzlawik et al.). Listening to a song again, for instance, is not the same since it involves our previous experience and associations when first heard and the present experience and associations. Recursivity, for the authors, is seen as a "looping-back" process that is related to self-awareness that occurs in a social system or context. Thus, the authors introduce a temporal dimension into recursivity that they then link to identity development. Next, they examine the conceptual and often thorny terrain of *identity* and *identity development.* The authors argue that identity is a not a static or singular concept; rather, instead of *identity* we need to take about *identities* and, importantly, we need to be aware that both the object (self) and the subject looking at the object are dynamic and constantly changing. Identity refers not to objective sameness but to the attempt to construct a semblance of continuity or something fixed onto something dynamic. Thus, identity refers to a structure of interdependent parts and a dynamic process of change. Moreover, this process is socially organized and culturally guided through the various frameworks provided in our various social milieus. However, as the authors argue, after reviewing the literature on identity research, they find that research on identity exploration has been present- or future-oriented. In other words, researchers and models of identity have explored how individuals compare their identity with their current situation and what they want to become. The authors, while not neglecting the present or future, emphasize the recursive process of "looping-back" to revisit the past in identity exploration:

> Recursion within, or better, as part of identity development is characterized, thus, not only as a forward-oriented endeavor, but also a process that includes looking back: The individual experiences something and evaluates this experience by (1) looping-back as the primary means of self-awareness (self-referencing); then, if the experience is not repelled or ignored, (2) change can be observed. As a third step, the new experience is (3) integrated and becomes meaningful for identity.... In this sense, recursivity is doubly appropriate: Individuals loop back, as a matter of process, and individuals look back, as a matter of exploration. (p. 171)

The authors conclude their chapter with three rich case studies to flesh out how individuals revisit the past in order to make sense of ruptures and changes in their lives and are changed through this process. Whether it is a woman examining her transition from a very religious to less religious individual, teenagers exploring their sexual identity, or individuals in the therapeutic setting exploring their problems, there are common recursive threads throughout each: The past is interpreted and integrated into the present as the individuals "move forward" in their life course.

Next, Luis Rivera, Ana Prados, Sandra Londoño and Mauricio Cortes (Chapter 7) examine how recursivity permeates the ancient texts of St. Ignatius of Loyola. St. Ignatius was a Spanish Jesuit priest who wrote his *Spiritual Exercises* based on his conversion and spiritual experiences around 1541. Focusing on St. Ignatius and his spiritual exercises may seem like a curious choice for a volume about recursivity and psychology. However, St. Ignatius' writings bring to the foreground the relevance of spirituality for the *psyche*. Hence, St. Ignatius and this chapter are significant for psychology and are rooted in profound existential desires of connection and transformation. Indeed, the exercises of St. Ignatius are filled with prayers, admonitions, and methods for self-care, finding one's identity, and ultimately finding and becoming unified with God. Cortes et al. carefully examine these exercises and provide a novel interpretation of their (perhaps transcendent) nature and efficacy by illuminating and elaborating their recursive characteristics. The authors draw on two complementary forms of recursion: that of auto-similarity found in fractals and the recursive elements of language. On the one hand, the structure of the *Spiritual Exercises*, Cortes et al. argue, reflects a mathematical form of recursion where the component parts and elements of the system repeat themselves in similar (though not identical ways), which in turn lead to new and novel cases emerging from an initial, base case (p. 6). Moreover, the authors highlight how the exercises also entail a conversation or dialogue with God and this dialogue is inherently recursive. Dialogue can take place between different *real or imagined* interlocutors as an exchange of communicative signs through which emotions and actions are coordinated. Insightfully, Cortes et al. combine both notions of recursion (linguistic and mathematical) to illustrate how they are manifested in the *Spiritual Exercises* and how they lead to transformative experiences:

> This way of coming back on each exercise is reproduced until the subject achieves change, generates novelty, constructs meaning, as happens when an individual talks to another and the exchange stops when the parties have understood what the other wants to say or have opened a way to comprehend the other's point of view and arrive to something that may be given a sense of dialogue. For this reason, talking to God from the exercises is a recursive experience from a linguistic perspective.... Dialogue despite being structured and following a procedure in certain way inflexible, not necessary arrives at the same point; *it may create novelty at the end of the repetition.* (p. 209; emphasis added)

This chapter is an interesting exploration of personal transformation through dialogue and offers a distinct yet complementary approach to identity exploration and imagination.

Finally, Zachary Beckstead (Chapter 8) offers an exploration of recursivity by looking at a particular act of commemoration: roadside memorials and shrines. In particular, he is interested in investigating how these frequently simple and occasionally elaborate objects captivate and arrest our attention and what they communicate to those who create, maintain, and pass by these material artifacts. He notes that many signs, objects, and practices relating to death are sure to be experienced as disturbing and fascinating. So, what this chapter explores is how these memorials can pack an emotional punch in spite of (or perhaps because of) existing at the margins and even though they are mostly encountered quite fleetingly. He suggests that the roadside memorial and cross create connections between the past, present, and future and link the living and the dead through dynamic and recursive processes. It is typical to say that roadside memorials and crosses make grief manifest; more to the point, he argues, they make temporal and social relationships explicit and provide the context and ground for our emotional engagement and perspective taking. Because of the roadside memorial and its capacity to evoke and unite different perspectives and positions, human beings can enter into the imaginative realm and feel and contemplate the perspectives of the bereaved, the dead, and the witnesses of these objects.

Taken as a whole, this book takes a complex concept and uses it to illuminate a range of phenomena in the human and social sciences through the lens of recursivity. Moreover, the authors loop back to consider how their investigation support, challenge, and modify our current notions of recursivity. Ultimately, the authors in this book present a strong case for taking recursivity as a central concept in the architecture of the human mind, social relationships and methodological framework of social scientists.

REFERENCES

Bánréti, Z. (2010). Recursion in aphasia. *Clinical Linguistics & Phonetics, 24*(11), 906–914.

Chomsky, N. (1957). *Syntactic structures.* The Hague, The Netherlands: Mouton.

Clifford, P. A., Friesen, S., & Jardine, D. (2001, April). *The ontology of hope: Lessons from a child.* Paper presented at the 82nd Annual Meeting of the American Educational Research Association, Seattle, WA. [accessible online via: http://jan.ucc.nau.edu/~chaplx-p/docs/CliffordetalAERA2001Paper.pdf]

Corballis, M. C. (2011). *The recursive mind.* Princeton, NJ: Princeton University Press.

de Saussure, F. (2011). *Course in general linguistics* (W. Baskin, Trans.). New York, NY: Columbia University Press. (Original work published in 1916)

Descartes, R. (2003). Discourse on method. In E. S. Haldane & G. R. Thomson (trans.), *Discourse on method and meditations* (pp. 1–52). Mineola, NY: Dover. (Original work published in 1637)

Driesch, H. (1925). *The crisis in psychology.* Princeton, NJ: Princeton University Press.

Everett, D. (2005). Cultural constraints on grammar and cognition in Piraha. Another look at the design features of human language. *Current Anthropology, 46,* 621–634.

Goffman, E. (1963). *Stigma: Notes on the management of spoiled identity.* New York, NY: Prentice Hall.

Lear, N. (Producer) & Reiner, R. (Director). (1987). *The princess bride* [Motion picture]. United States of America: Act III Communications.

Perinat, A. (2007). Comparative development of communication: An evolutionary perspective. In J. Valsiner & A. Rosa (Eds.), *The Cambridge handbook of sociocultural psychology* (pp. 140–163). Cambridge, UK: Cambridge University Press.

Sawada, D., & Caley, M. T. (1993). Complementarity: A recursive revision appropriate to human science. *Anthropology of Consciousness, 4*(2), 1–8.

Vitale, B. (1989). Elusive recursion: A trip in a recursive land. *New Ideas in Psychology.* 7(3), 253–276.

von Humboldt, W. (1999). The diversity of human language-structure and its influence on the mental development of mankind. In P. Heath (trans.), *Wilhelm von Humboldt: On language* (pp. 1–287). Cambridge: Cambridge University Press. (Original work published in 1836)

CHAPTER 1

MULTIPLE PRESENCES OF RECURSIVITY

Adolfo Perinat
Universidad Autónoma de Barcelona, Spain

ABSTRACT

Recursivity is a capacity unique to the human mind. In the domain of linguistic discourse, it is evidenced by to and fro movements followed by the superimposition of a new direction and gives rise to paradoxical statements or the insertion of parenthetical statements—narrations within narrations. In the case of behavior, it takes the form of shifts between different scenarios created by the mind, such as play, rituals, or theater. In general, a fragment of a stream of behavior or a proposition is encapsulated, separated for a moment and then reincorporated. It is clearly possible to show that the underlying mental operation hinges on the unique human psychological capacity of being, at the same time, "inside" and "outside" the main stream of the mental process that guarantees the continuity and coherence of all human behavior. We characterize this capacity as becoming an observer. It is fully applied in the case of play and fantasy dreams, that is, in instances of the "other realities" that humans are capable of inhabiting. The final implication would be to raise the question of whether the particular aspects of the reality defined as play, theatre, ritual, myths, and so on, are part of our life as legitimately as the reality regarded in a vulgar sense as "real."

Cultural Psychology of Recursive Processes, pages 1–21

INTRODUCTION

The term *recursivity* comes from the Latin *recurrere* (to go back or return). The primitive meaning of the term was "to retrace the same path." Thus, it implies a return to the beginning and the repetition of the same trajectory over again. We find the same original meaning in the term *reiterate*. However, while the root of reiterate is *iter* (path or journey), as a result of a metaphoric displacement the term is applied to linguistic discourse. Recursivity or, if you prefer, recursive processes are manifestations of human cognitive activity characterized, in a first rough approximation, by the fact that the mind's discourse executes a back and forth movement or makes "jumps" between the different domains in which it operates. These movements can result in veritable paradoxes. For example, the criticism of art or fine arts is recursive when it responds to criticism. Also recursive is the appreciation of the plays of Samuel Beckett that express the notion "*Rien n'est plus essentiel que l'inessentiel*." The recursivity in such cases is revealed by the fact that the discourse loops back, resumes its initial direction, and apparently contradicts itself: criticism criticizes itself; the essential lies in what is not essential. In these twists in the discourse that turn back upon themselves to then resume a parallel course, circumventing the principle of non-contradiction, we find the "hard core" of recursivity. Another instance of recursivity, which also takes the form of a back-and-forth movement or jump, occurs when we open a parenthesis or place a footnote on a page: the discourse is briefly interrupted and then immediately resumed. The notion of the footnote opens the door to a productive expansion of the train of thought that takes us to the concept of "as if" play. "As if" or make believe play also involves a jump between one domain of reality and a separate scenario created outside of that domain. However, the existence of two scenarios (one commonly called fiction and its counterpart reality) implies a border separating the two. The parentheses in the discourse also imply a break or pause. One of the key points in this chapter is the proposition that the recursive mind crosses borders; it is a "smuggler" that moves with ease between territories. And, since we are creating scenarios and drawing borders, why not analyze theater, ritual, therapeutic conversation, the anthropologist's field work, and other human situations as instances of recursivity?

The purpose of this chapter, as the reader may already have guessed, is to show that there are (discursive) mental acts, such as reiteration, shuttling between mental domains (back and forth), drawing borders between domains, creating scenarios (called fictitious), and so on, that are recursive in nature. Recursivity has its roots in a mysterious ability that emerged during our evolution. Mysterious? Yes, because, as we will argue, to state a paradox or to dilute an inherent contradiction and draw forth its intelligibility is a process born, we know not how, in the unfathomable depths of the human mind. Reasoning about recursivity is a complicated process about a process. What we can surmise is that our recursive capacities were born with language. To put it more precisely, recursivity emerged and made language possible at the same time that language emerged and increased our recursive capacities. (This itself is a recursive process: autopoiesis, the key to emergent phenomena.) At this point, I will pause for a moment, hoping that the reader, after this arduous preamble, will feel ready to start out on the long and tortuous path of "initiation" to glimpse at the end the radiant light.

THE EMERGENCE OF THE OBSERVER

To approach the notion of recursivity I am going to start with a general framework in which I will present a character we shall call the *observer.* The notion of the observer as proposed here goes beyond the trivial meaning of the word as someone who merely looks on, recording what he or she sees. In this argument, being an observer or becoming an observer refers to a mind that not only sees phenomena but also comprehends (attributes meaning to) the phenomena observed. This is a typically and exclusively human ability: Animals can look and can see, but they cannot be observers. The figure of the observer in our research emerged first as a real person and later acquired a metaphorical sense, that of a "way of seeing" human actions.

In various studies undertaken in the 1990s by our research team (Lalueza, 1991; Lalueza & Perinat, 1994; Sadurní & Perinat, 1994, 1995, 1999; Sadurní, 1992), we filmed the interaction between children and their mothers at play. Two episodes were particularly memorable. In the first, the mother and her 28-month-old child Jaime played "eating soup" using real spoons and imaginary plates. Jaime's mother lifted the empty spoon to her child's mouth while cautioning him, "Careful now! Don't burn yourself! Blow on it first." Jaime dutifully blew on his "soup" to cool it and opened his mouth to allow his mother to "feed him." He pretended to eat the soup and smiled at his mother. Immediately after he had "eaten his soup," Jaime turned and looked directly at the person filming the scene with a smile of complicity. The second episode involved Ana, a 24-month-old girl. She and her mother were playing with a *porrón.*[1] When her mother positioned the *porrón* close to her mouth signaling her intention to "drink" from it, the little girl—glancing in the direction of the researcher filming the scene—said, "Mama, what will Marta say if you drink wine? "The children, both with a look and Ana using a direct allusion, made it clear that they were aware of being observed. Someone is standing outside the scene—the observer—a person whose gaze and whose appraisal of what is happening has started to have an effect on their actions. The key point is how, from this moment on, the situation and the people involved in it are altered. Let us take a closer look at this shift.

Common sense would suggest, at first glance, that the observer stands outside and is separated from the scene he or she observes. The researcher filming and the child playing are operating in different domains. However, when the children engage the gaze of the observer with their eyes, they draw the observer-researcher into their play scene. They involve the observer and at the same time they perceive themselves to be involved by the observer. The little boy creates the observer when he perceives himself to be observed; ipso facto he himself becomes the observer of the observer. This is an interesting instance of *reflection. (I will use the notation *reflection to indicate that I am referring to the physical phenomenon of mirror image reflection and not the intellective process). The child sees the observer and thereby becomes an observer. This is a self-referential act: Only an observer can see the other who is outside as an observer. In the children's first smiles of complicity directed towards those who are witnessing their shows or "as if" play, we sense the awakening of their observer's mind. However—and here is the superb paradox—the (human) observer is someone who, as we have seen above, is involved in the situation they observe. The observer is "dragged" into the situation by the complicit gaze of the children. This is something that can only happen if there is an "expansion" of the scene, which from this point on includes the new character. (Otherwise, what sense would there be in involving the observer, even if only with a look?)

In classical cosmogonies and theogonies, the gods look down from their Empyrean heights and observe what is happening in the world. In the world, humans struggle and fight under the indulgent or punitive gaze of the gods. The human sciences have rebelled against the gods by creating the figure of the observer, a figure who has moved beyond this state of estrangement ("Ye shall be as gods!"). In this category we find the anthropologist, the sociologist, the therapist: in short, all those who not only observe human action but also participate in the same human action they observe. They wrestle in the antinomy of not being a participant and participating. This is one of the central points in any discussion of this situation: the fact that, unlike the gods, we mortals are at once the observers and the protagonists of human action. In any case, the tragedy of Paradise (the paradise of achieving an always elusive complete and perfect knowledge) is something that we in the social sciences relive constantly in our struggle to explain our own behavior in objective terms—as outside observers—and at the same time to understand it in a meaningful way from the point of view of the other (*Verstehen*). In other words, we are observers in a domain of shared meanings and experiences, which we use to explain ourselves. Because, ultimately (and this is the compelling corollary), a person who does not make use of his capacity to observe himself cannot be called an observer in the strict sense of the term. The paradox this generates (that of the observer who observes himself) and its cognitive resolution lie at the center of our ability to play, to symbolize, to fantasize, to joke, to create rituals, to play roles; they are also fundamental to the theory of mind and many other ordinary behaviors that are no less puzzling for being so commonplace.

Ever since the mid-19th century when psychology first started to jostle for a place among the sciences, striving to create a respectable domain for the discipline, there has been a split in the way psychologists approach the phenomena of the mind or psyche (better known at that time by its former name, the "soul"). The study of the physiology of the nervous system opened up unsuspected vistas in our knowledge of the sensations, but how to become aware that you are aware of (and describe) what you feel? What should we do with the phenomenon of consciousness (and, in passing, that of introspection)? Mechanicists, materialists, reflexologists, and neurophysiologists comprised the band of those who advocated objectivity or detachment in the study of the mind, treating it as a "natural" phenomenon and using methods as close as possible to those of physics. In the opposing camp, we find those who defend the specificity of the social sciences with Vico as an 18th-century precursor of this current. The posture of this camp eventually cystallized in the concept of *Verstehen*: the experiential understanding of the experience of the other. Ever since, the schism has persisted and the controversy has raged unabated. Likewise, we continue to give an account of human behavior not in terms of an "objective" description (like that of the path of a planet in its orbit or of the water in a river bed) but rather as actions set in motion (motivated) by intentions and purposes. In other words, we not only describe human actions, but we *explain* them and attach social meaning to them. Looking at it this way, we could say that we make *semantic descriptions* of behaviors.

To conclude this first approximation, we can clearly define that the observer is *a person capable of making semantic descriptions of the behavior of an actor or actors*. To put it inversely: we humans are innate observers *because* we spontaneously make semantic descriptions of human actors. And not only of human actors: How often do we produce these types of descriptions of domestic or wild animals? And what else is the animism (not only Piaget's infantile type) that we discover at every turn in behaviors that the "enlightened" could not fail to qualify as irrational? I will add that, although an observer can make "objective" descriptions and treat the actors as inanimate beings directed solely by physical causality, he can only do so by exercising a great deal of imagination, and he is always aware that he is making a conscious effort to avoid semantic description.

THE OBSERVING MIND

At this point we will delve a little deeper into the concept of the observer or observing mind. Let us imagine that we are the spectators of a film or imagine ourselves in the situation of a person attending a ceremony belonging to an exotic culture—for example, an African ritual. In both of these situations, there is action on one hand and a person viewing it on the other. Inevitably, however, at some point in the film, anticipating that the protagonist is about to fall into a trap, the spectator will feel moved to intervene (by introducing himself into the scene in his imagination) in order to warn the character what his enemies are plotting or simply berate him for his lack of caution. The spectator was outside but is also inside. The same thing can happen in the case of the person attending the African ritual. Perhaps at a given moment the spectator may feel spontaneously compelled to take part in the ritual or he may allow himself to be carried away by the magic of the ceremony or may be invited to participate. In neither of these two different scenarios (the film or the ritual) does the spectator ever lose his awareness of the fact that he is outside of or alien to the situation, but in both cases he nonetheless feels immersed in it. The same thing happened in the episodes described above involving the children Jaime and Ana.

In all of these cases there are two clearly distinguished moments. In the beginning there is an onlooker who observes. Before him, in a domain that he sets apart and delineates, are the actors and their drama. In the case of the African ritual, the foreign visitor may have virtually no idea of what the ritual means; he merely sees movements and gestures (an objective description). But there comes a time when, while remaining an onlooker, he also becomes another actor and starts to take part in the action, at which time the movements and gestures begin to take on some meaning because of his involvement in a group dynamic. (He is still ignorant of the "higher" symbolism involved, the allusion to ancestral myths and so on.) But this vague idea is not enough; when, at the end of the ritual, the dancers explain the meaning conveyed by their movements and gestures, only then will his participation in the ritual start to make sense to him. In this example (which could be extrapolated, for example, to the therapeutic relationship), we can see that becoming an observer is a never-ending process: There are always new layers of meaning to uncover in human action. There are two points I would like to emphasize here. First, it is important to note that the border dividing the two domains has been eliminated: There are no longer two domains but only one in which the actions of the newly incorporated actor acquire the meaning that has always been attributed to them by the original participants. This new domain is an extension of the original domain. Second, and also important, is the fact that the ultimate meaning of the ceremonial ritual is revealed through language. This is much more than a detached description: It is a description that takes on meaning when comprehended in conjunction with the action observed.

The breakdown or expansion of borders is an example of "Gödelization" (an application by analogy of Gödel's theorem)[2]: The semantic descriptions that struggled to emerge from the observer's mind were finally introduced into the domain of observed behaviors. What is the implication of this operation? Maturana and Varela (1990) provide the answer: The system of "operational descriptions" (the one used by the person looking at the African ritual who only "saw" movements and gestures) cannot give an account of certain phenomena in the behavioral domain. To put it another way, it is not acceptable to limit oneself to *describing* behavior; it is necessary to *explain* it. Gödelization creates (for the observer) a class of behaviors that were previously nonexistent: sign-behaviors. The person observing the African ritual desperately sought to interpret the gestures and movements of the dancers but lacked the semantic keys to interpret them. His communion with the dancers opened an (analog) way to a basic understanding; the explanation he later receives *makes those behaviors become signs*—that is, it endows them with meaning. Once revealed, these "new" behaviors occupy the place of all the previous ones, and there are no longer behaviors for the onlooker and signs for those taking part in the ritual, but only a single class of significative (meaningful) behaviors.

The fusion of signs and behaviors leads us to consider behaviors as signs and can be represented as a play of mirrors (*reflection)

behavior-sign] [sign-behavior

that must be interpreted as "the behaviors seek the sign" to take on meaning and "signs seek the behaviors" to endow them with meaning.

I have tried to show that for the outside observer to become an observing mind it is necessary to eliminate the border, a step that implies an expansion of the domain (Gödelization). In taking this step we have been helped by the notion of *reflection. I will now show that rather than being peripheral to the process of becoming an observer *reflection lies at the heart of that process. To illustrate this idea, I will use an example-anecdote. In a large tourist destination (I write this in Barcelona, Spain), a man strolls along a wide avenue crowded with people. A mischievous clown, initially unseen by the walker, follows closely in his footsteps, mimicking his gait. When the man turns around suddenly the clown stops for an instant (surprised?) and then starts to back away from the man with tiny mincing steps. The walker then follows in the clown's footsteps, mimicking his mincing steps; the person being imitated has become the imitator. His reaction is a mirror reflection of the clown's behavior. The walker and the clown now both form part of the same show.

*REFLECTION, SELF-REFERENCE AND RECURSIVITY

The walker who turns around and, retracing his own steps, replicates the clown's imitation of him performs a recursive action analogous to the reflection of a beam of light by a mirror. Language lends itself spectacularly to such reflective plays. *Reflection lies at the root of a series of paradoxical statements that, while well known, never lose their ability to shock when we reflect (without *) on them. I will give some examples:

> "I only know that I know nothing" (Socrates)
> "Il est interdit d'interdire" (It is forbidden to forbid); slogan coined in May 1968 in Paris.
> "All Cretans are liars," says Epimenides of Crete.

The examples are myriad. It is as if the notions of *knowing, prohibiting, lying* in the very act of being proclaimed, collide with an unseen wall that reflects them back as they contradict themselves: "I know I do not know," "Forbidding is forbidden," "I, lying Cretan, affirm that Cretans lie." The end point unexpectedly bounces back to the starting point by way of a curious trick: the loop. And the final result is a closed circle. Despite the logical incongruence, the sentence makes sense. This strange situation to which we have come could have been anticipated from the beginning of this chapter from the examples of the criticism of criticism and the essential nature of the nonessential.

Such discursive constructions are typically self-referential. This is clearly evidenced in knowing, prohibiting, lying, criticizing, and in what is essential: The utterer and the utterance are inappropriately associated; the concept predicated is the same as the concept that organizes the predication. The same self-referential quality is also found in the processes of academic reflection labeled with the prefix "meta": metalanguage, metacognition, metahistory, and so on. In this case, there is no contradiction but rather a duality of logical planes: the language that talks about language, knowledge of how we know, a story about what it is and how it has become history, and so on. This dual logical plane is also revealed in the (paradoxical) statement made by the Cretan Epimenides and also in other statements, such as:

> Qui custodiat ipsos custodies? (Who will guard the guardians?)
> Learning to learn
> Never say "Never"!
> Groucho Marx: "I'd never join a club that would accept members like I."

At this midway point along our route, I would like to return to the thread that has led us from the figure of the observer to the dissolution of borders, the emergence of significative behaviors (associated with the metaphor of *reflection) and the notion of self-reference. And where is the recursivity in all these cases?, more than one of you will ask. For one thing, self-reference is in itself a *recursive loop*: the ideas of guarding the guards, learning to learn, never saying never, and not joining a club that would accept one as a member all involve returning (in one's mind) to the beginning of the statement and retracing the same path a second time. This notion of *return* can be understood through the analogy of light reflected in a mirror. If at the *terminus a quo* we place a second mirror, the phenomenon of *reflection will be multiplied ad infinitum. Nothing illustrates this better than the phrase, "I think you think I think you think," which is an example of pure recursion.[3] But let us go back to our starting point. To be an observer, I said, is "to distance oneself," "to delineate a domain" (actions and actors), "to make semantic descriptions." If being an observer and being an actor were mutually exclusive states, the discussion would end here. Recursivity is a mental construction that allows us to be both observer and actor at the same time. What I have called "jumping outside" is nothing more than a spatial metaphor that describes the cognitive act of observing oneself as an actor (1st loop) and observing oneself as the observer of oneself as an actor (2nd loop), and so on *ad infinitum*. The analogy of *reflection provides the key to the relationship between "behaviors" and significative behaviors, the latter being behaviors that have been "discovered" by the observer. *Reflection is also the key to the link between recursivity and self-reference.

RECURSIVITY AND FRAMES

We find another important aspect of recursivity in discourse: the use of interpolated clauses, parentheses, excursus, footnotes, and relative clauses. This aspect is dealt with very amusingly in Hofstadter's *Gödel, Escher, Bach* (1979), a book from which I have taken some of these ideas. In this case, the recursivity takes the form of a "descent" to another level for a short period and a subsequent return to the original level at exactly the point of departure. A person reading aloud, after reading a footnote, announces to the audience, "The text continues." In this case, the recursivity takes the form of embedding and is, as Hofstadter is pleased to demonstrate, a process intrinsic to linguistic discourse. One of the earliest manifestations of recursive thought in children can be found in metalinguistic comments (statements about statements made by another person).

The notion of recursivity as embedding or parenthesis can serve as a "visual" introduction to the subject of boundaries or what in the title of this section I have called frames. I have taken this term from Gregory Bateson as a tribute to his excellent work "A Theory of Play and Fantasy" (1955/1972), a paper that has inspired many of the ideas discussed below. A recursive loop—I propose—establishes a frame; in other words, it defines a subset of the main set. However, at the same time, the recursive loop maintains a connection between the framed subspace and the original space within which the frame has been drawn.

What I find most interesting in Bateson's reasoning is the shift he proposes from logical frames (Epimenides' paradox) to psychological frames. His thesis is that play, ritual, fantasy, and other typically human manifestations are only possible if we establish a frame (a parenthesis or embedding, we would now say) that borders on the paradoxical but which—given its psychological rather than logical character—is perfectly admissible to the human mind. Describing the play of two monkeys at the zoo, during which the primates chase and "bite" each other, Bateson notes that the play involves a paradoxical frame, which he describes in the following way: The play actions in which they are engaged "do not denote what those actions *for which they stand* would denote. The playful nip denotes the bite, but it does not denote what would be denoted by the bite" (Bateson, 1955/1972, p. 177, emphasis added).

The reader will easily make two connections: one with recursivity and the other with the condition of the existence and use of symbols. Bateson himself reveals this recursivity/*reflection when he goes on to say that "the word 'denote' is being used in two degrees of abstraction, and these two uses are treated as synonymous." "It denotes such a thing" and "It denotes [that denotes such a thing] ." The two levels are conflated. To the logician's *non sequitur,* the psychologist's gleeful retort is *Eppur si muove!* Play exists; jokes exist; fantasy, theatre, ritual, and symbols all exist. Recourse to the well-known distinction between the map and the territory allows Bateson (and us) to breach the wall of logical incongruence and penetrate the realm of the mind and psychological phenomena, which is, par excellence, the realm of ambivalence.

Allow me to illustrate my idea with the light-hearted topic of jokes. Bring to your mind one of the best jokes you have ever heard. The joke teller draws a map that suggests a territory, but one that we know from the beginning is not the actual territory that corresponds to the map in question. The climax occurs when the punch line reveals the real territory: the more unlikely and unexpected the denouement, the funnier the joke. In this sense, a joke represents a kind of trance (or rupture). Before we move on to deal with the subject of play, I would like to call attention to some points that have started to become clear: first, that play involves a typical case of a boundary that frames the participants (in the joke or the play); second, that play is governed by the distinction between the map and the territory but not by a distinction that entails the either/or exclusion imposed by logic but rather by one etched in the ambiguity and ambivalence of psychology.

Finally, I will show that only the observer can establish the boundary (frame). Establishing a frame is analogous to defining a subset within a set. If the new subset is delimited within an existing set, one is acting from outside (with respect to the subset); if it is established through an expansion in which the original set ends up being encompassed by the larger set, one is acting from inside. In the first case, the person delineating the frame was previously the observer and he acts by delimiting; in the second case, the person delineating the frame becomes an observer when he frames (himself). The same thing occurs when the frame is broken and the boundary eliminated. The outside cannot exist without the inside and vice versa. And the observer is the only one who can create that relationship.

In short, the recursive itinerary can be represented by the "circuit" in Figure 1.1.

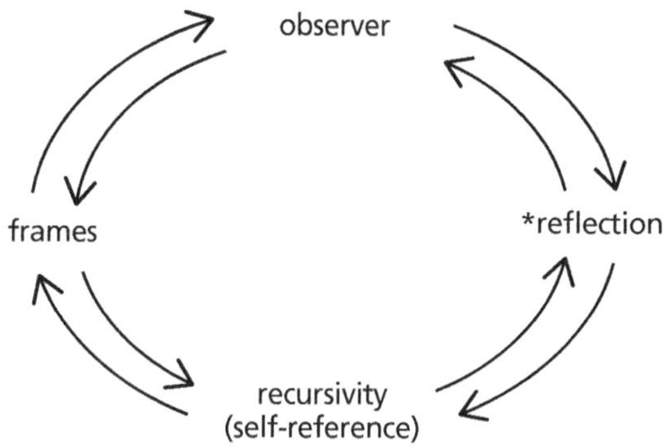

Figure 1.1

This is what Hofstadter (1979) calls a *strange loop*; it occurs when, after moving up or down through the levels of a hierarchical system, we unexpectedly find ourselves back at our point of departure. In other words, we have executed a recursive loop. (Introducing recursivity, I myself am caught up in the snare of a recursive loop!)

"AS IF" PLAY (MAKE BELIEVE)

In the article cited above, Bateson affords us a magnificent avenue of entry to understanding play: play involves establishing a frame; play is a map and therefore refers to a territory. Among the string of implications we have discussed above, play involves recursivity, *reflection, and becoming an observer. However, although the formalism of the frame and of the *reflections/recursivities brings us closer to play, the essence of play lies in the psychological, and this is where we need to investigate further.

In the naive conception of play proposed by folk psychology, the assumption is that there is a reality external to the play situation and that the player imports this reality into the play "territory." It is accepted that play involves establishing a Batesonian frame ("Let us play"), but at the same time it is assumed that what happens inside the frame is a (more or less distorted) reproduction of real-life situations. Piaget and other authors make reference, technically, to "decontextualization"; Leslie (1987) proposed the operation of "decoupling" and introduced the concept of "metarepresentation," but he did not infer all the consequences he might have from this (implicit) invocation of recursivity. The paradox aptly stated by Bateson is maintained but not overcome: "It does not denote what would be denoted," because, by accepting the mutual exclusivity of reality and fiction, they religiously follow the logical principle of noncontradiction and therefore do not feel the need to circumvent the paradox. Nor do they apprehend the status and position of the observer—the person who establishes the frame—nor the back-and-forth of *reflection and recursivity that this entails.

Let us start by clarifying the psychological status of the observer. I will make two connected proposals here. The first is that the observer is someone who moves from inside to outside and vice versa with astonishing psychological flexibility, a flexibility that is so common to humans that no cognitivists have posed the problems involved. In our observations of mothers and children at play, the mothers "jump" out of their play character with absolute mastery and spontaneity to instruct the children how to follow the script of the play (see, for example, Perinat & Sadurní, 1995). You might say that the mother shifts from being a participant to being an observer and vice versa. True, but there is more to it: The participant and the observer are not two different entities but rather two different states of mind that the mother handles at the same time. The participant does not stop being an observer; if she did, how could she tell the child "you are not doing it right" or explain "that is not meaningful behavior"? Hence my second proposal, acceptable from the standpoint of psychology and unacceptable from the standpoint of logic: The observer is at once inside and outside the frame. The more involved the observer becomes in the play, the more she lives the fiction: She is immersed in the feelings of the characters and she plays her role splendidly. (We have finally come to theater!) Only the observer-actor can shatter the paradox; only the observer-actor can reconcile the opposing poles of fiction and reality.[4]

These two proposals closely follow the inspiration of Bateson, who warns us that logical frames (those that generate paradoxes of the Epimenides or Quine[5] type or Russell's paradox, so well known to mathematicians, which inspired Gödel's theorem) are radically different from psychological frames. "It would be bad natural history to expect the mental processes and communicative habits of mammals [and a fortiori of humans] to conform to the logician's ideal" (Bateson, 1955/1972, p. 180) The same idea is expressed by Varela (1989) when he says that Epimenides-Quine type phrases explicitly violate the assumption that "what we say about something should not enter into the constitution of that something." The two levels of meaning of the phrase should not intercross; if this condition is not met, the phrase is meaningless.

> It is harder to leap out of the need to stay at a given level of meaning and simply look at the whole sentence as a unity.... In the Quine or Epimenides case, the phrases remain a paradox unless I am willing to let go of the need to choose between true or false, and see the sentence's circularity as its own way of specifying its meaning. (Varela, 1989, p. 23)

If we transpose this idea to the domain that concerns us here, the person playing becomes involved in a circular process that results in the dissolution of the boundaries (frames) in her mind. She does not know (or care) whether she is inside or outside the frame; she conflates the two. In short, going beyond what in a logical analysis is a case of two topologically distinct instances, the mind has the ability to pass effortlessly from one plane to another and in the end to simultaneously maintain, to whatever degree it wants, both the frame and the dissolution of boundaries, both the distinction and the conflation of territories. We can be both observer and participant at the same time; we can handle the recursive twists of the loop; we can travel on the wings of what is *reflected; we live in an interwoven duality.

On the subject of the strange loops characteristic of Escher's work, such as his *Drawing Hands* (Figure 1.2), *Waterfall*, and *Print Gallery* (Figure 1.3), Hofstadter writes:

> In some of his drawings, one single theme can appear on different levels of reality. For instance, one level in a drawing might clearly be recognizable as

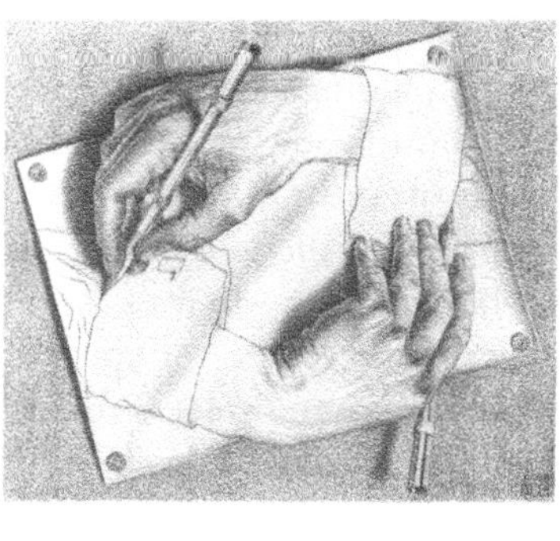

Figure 1.2 Drawing hands (Escher).

Figure 1.3 Print gallery (Escher).

> representing fantasy or imagination; another level would be recognizable as reality. These two levels might be the only explicitly portrayed levels. But the mere presence of these two levels invites the viewer to look upon himself as part of yet another level; and by taking that step, the viewer cannot help getting caught up in Escher's implied chain of levels, in which, for any one level there is another level above it of greater "reality," and likewise, there is always a level below "more imaginary" than it is. This can be mind-boggling in itself. However, what happens if the chain of levels is not linear but forms a loop? What is the real, then, and what is fantasy? (Hofstadter, 1979, p. 15)

This puzzling question is what interests me here. What is reality? What relation does play have to reality? Because this unappealable contraposition of fiction and reality lies at the heart of theories about play. The child and the adult play "dinner" (with toy cutlery), "trains" (with wooden blocks that clip together), "phones" (with a toy phone), and "putting baby to bed" with a doll and some wipes. And just in case anyone feels that these examples all belong to a bygone age of traditional scenarios, we can also add the example of virtual games in which somewhat older children manipulate characters and follow (or direct) them in lifelike situations within a virtual reality. In both cases, the children are acting out "as if" situations. In any case—we are told—they are building a fiction based on reality. I believe that we need to seriously question this way of seeing things. The notion of the frame, with all it implies, invites us rather to consider—as Hofstadter suggests—not the "reality/not reality dichotomy but rather *different hierarchies of reality*"—in this case, what happens when we're playing and what constitutes "life." The jump between these different orders of reality (or the location of "what happens" on different levels) is the subject of strange loops. The essence of play is that we are at first aware of this duality; however, in the course of play, this awareness fades and we confuse the "here and now" of play with another "here and now" that is not play.[6]

Despite appearances, for the child, play does not initially represent an "as if" situation for the simple reason that children do not and cannot distinguish between the hierarchical levels of reality. The idea that different levels of reality exist has to first develop in their consciousness. In other words, the child has to become an observer. When the little girl plays "dinner," "trains," "phones," or "putting baby to bed," the play is symbolic (in the traditional sense of the term) if and only if she does what she does while remaining aware (virtually) of the other level. What she does is a product of the fusion of the two levels. She is obviously making use of schemes of action, and some of these schemes are the same as those adults use in "real reality"; what I reject is the notion that it is clear in the child's mind that she is importing them from that other reality. They are clumsy imitations, and she lacks a clear awareness that they constitute (adult) reality. What I believe is that the child wakes to life as does the solar day—in the half light; the distinction between what adults call play and what they call real life is something that dawns on children as they develop cognitively in a domain of typically communicative interactions. And this is true not only of the "frames" of play but also of many other situations of daily life: The mother uses play to encourage her child to eat, when she bathes him, when she puts him to bed, tells him a story, and so on. Adults make constant use of fantasy (or what they call fantasy).

I said earlier that play involves conflating the map and the territory while at the same time keeping in mind the distinction between the two. It might be better to say that it involves *making a pact with the other on the distinction.* This is what makes us intuit complicity in the look that the children directed at the person who was filming them from outside the scene in the anecdotes I cited at the beginning of this chapter. Not only did they *see* a character who was filming, but they also *became aware* of her appraising observation of the scene. They went on to incorporate this character into the frame; they eliminated the border but continued to play.

From that point on, the play consisted of a double pact:

1. The map is not the territory.
2. We deny that "the map is not the territory."

In other words, we deny (1). Thus, we are stating that "the map is the territory." We eliminate the logical line separating the two; we agree to suppress the frame. Note that we are now in the same conceptual ground as that of the Epimenides paradoxes and those we used to introduce the notions of self-reference, *reflection, and recursivity.

Condition (1) sets the context or frame of play. Condition (2) allows the use of the symbols belonging to the territory; it also allows us to confuse or distinguish the map and the territory without indexicalizing when we pass from one to the other. Note that for the pact concerning the suppression of the frame to be effective, the child must first distinguish between the map and the territory. (You can only agree to eliminate boundaries that clearly exist.) All of this is possible because of the pristine lack of distinction between behaviors and signs or because there are no longer pure behaviors but only signs. This is the condition of the observer.

This proposal is not incompatible with the notion of decontextualization. It assumes decontextualization, but within a much broader frame that also eliminates its great a priori—the opposition between reality and play. It also calls into question the hypothesis that the child is behaving "as if" when the adult observer decides that he is behaving "as if." If we accept the existence of multiple hierarchies of reality, decontextualization becomes transcontextualization.

* * *

If play is a scenario, an action involving actors who are also observers, this structure can be transposed to other scenarios in social life. The most immediate is the one I suggested a moment ago: the theater. And from theater to ritual is just a step since rituals are a (theatrical) representation of a memorable event (religious services, annual carnival celebrations), a ceremony paying homage to a person (investiture of a ruler, Nobel Prize awards), rites of passage (weddings, college graduations), festivals (bullfights, circus), sports competitions, and so on. The ritual involves a frame that is equivalent to "this is play": It is not reality. But it does refer to some sphere of social life, which it reaffirms by evoking; the function of the ritual is to strengthen social bonds. Rituals are also governed by the distinction between observers and participants. The border between the two spheres is more or less stable. I alluded above to the outsider who attends an exotic ritual (African rituals are quintessentially exotic, but those of the Masons or those imposed by the sorcerer Don Juan on the anthropologist Carlos Castaneda are no less so). With that story I made the point that the outsider started out as an observer and ended up as a participant. This is particularly common in rituals that favor trance states. The person who goes into a trance crosses the border and enters into a new territory; he does not return. Other kinds of rituals maintain a clearer separation between those officiating and the public, but, in any event, attendees who are overcome by emotion because of what they see can be said to be "inside" and in communion with whatever the scene represents (the Greek catharsis). This formal structure involving a scene and an audience (observers) is found with many variations in social life: the lecture, the therapeutic interview, the political rally, student protests. All social life is clothed in and governed by rituals, as has been clearly demonstrated by social scientists such as Clifford Geertz,

Victor Turner, and Ervin Goffman. The central idea in the present chapter is that this means that the human mind is also constantly managing recursivity in this "to and fro" shift between frames.

RECURSIVITY AND LANGUAGE

At the beginning of this chapter I alluded to the key role played by language in the genesis of recursivity. Language is a fundamental component in this discussion and one that deserves extensive development. However, here I will add only a few comments. Language is always talking about itself. Any statement inextricably entwines a reference to extralinguistic entities and a reference to language itself (the intralinguistic context). In other words, everything that comes to us through linguistic signs we apprehend at the same time as we apprehend the linguistic signs.

Recursivity—the capacity that developed in the primitive mind—played a key role in the origins of human language. Many studies on the ability of chimpanzees to learn language have sought to reproduce, on a proportional scale, the conditions that led to the emergence and development of language in our species. I will use as a reference here the well known work of the Rumbaughs (Rumbaugh, 1977; Savage-Rumbaugh, 1986; Savage-Rumbaugh, Murphy, Sevcic, Brakke, Williams, & Rumbaugh, 1993). During the initial phase of training, the chimpanzees in these studies learned to associate signs (lexigrams) with different food items. Later they learned the signs for giving or requesting something. They were then trained to construct noun–verb "sentences" to request food items. But instead of well-formed sentences, they produced chaotic combinations of lexigrams. Undefeated, the Rumbaughs trained the chimpanzees to eliminate the forbidden combinations from their "language." Finally, the animals learned that, in addition to the primary immediate relationship between lexigrams and food items, that there were other relationships of a higher order between the lexigrams (functionally "words") that governed how the lexigrams could be combined. The chimpanzees, in sharp contrast to what happens with human infants, were incapable of inducing these higher-order relationships themselves. Nonetheless, their achievement, even with a very limited set of token-words, was an impressive intellectual feat. What underpins this logical-cognitive operation of combination and exclusion that children perform from a very early age when they construct sentences? The key is that the signs do not "function" only in reference to a given object or action, but that there is a higher-order relationship between them that governs how they can be combined. A linguistic utterance involves two concurrent operations at two different levels: the relationship of the words to what they mean (the sign and its meaning) and the relationship of the words to each other (the grammatical structure). (To achieve the grammar it was necessary to create words devoid of extralinguistic references.) Deciphering this second order relationship is a *recursive operation* because the (human) mind must discover a *relationship between relationships.* This is a frontier closed to nonhuman minds. Or, to put it another way, it is a frontier that primitive (or not so primitive) minds crossed one day in the course of their evolution.

The above analysis will help us to get closer to the operation (also recursive) involved in the construction of sentences. Sentences are generated from a mental representation with an ideational and potentially linguistic content. As the string of words that will express that content unfolds, the linguistic subject must keep in mind a number of levels: the words themselves, their grammatical relationships, and the syntax that governs their ordering. This intertwining of levels should lead to a point at which meaning and structure are fused into a higher level: the phrase. Language is therefore a beautiful example of an autopoietic process (Maturana & Varela, 1980, 1990). The term *autopoiesis* is a compound word made up of *auto* (oneself) and *poiesis* (creation). It is a phenomenon typical of all emergent processes of self-organization, and its key feature is that the process and the product are on the same level, a feature clearly evident in language. However, the same thing can be stated by saying that the products of the process are fed back into the process, leading to higher levels of complexity. Once again, language is a clear example of this type of process: It self-organizes as it is generated. Language is the paradigm of *reflection/self-reference/recursivity.

Once we perceive the equivalence between autopoietic and emergent processes, the concept of recursivity (their *modus operandi*) acquires a stunning breadth. Emergent processes occur in the generation of living creatures; during human evolution they have flowered in many areas affecting the capacities of the mind. For example, and still staying with the subject of language, many years ago Stern described how children one day discover that everything has a name. At around two years of age, the child's name-generating capacity appears. But the act of naming (as a conceptual operation) is "to name the act of naming." In the phylogeny of our species back in the mists of time, another creation of enormous social significance also emerged: myths. Merlin Donald (1991) has suggested that the most primitive forms of language came into being to express the myths of those ancient societies. Maybe those myths were expressed in their pristine simplicity, by mimed actions: and that may have been how ritual was born. As Northop Frye (1990) says, the myth is a story that, on the one hand, explicitly says, "This happened in this way," but also implicitly says, "This could not have happened in this way." The same thing happens, in a similar way, with metaphors: for example, "Your laughter is a sudden silvery wave." No one in their right mind thinks that stories such as those of Prometheus, Lohengrin, Frankenstein, or the Sorcerer's Apprentice are true or thinks that a girl's laughter is actually a wave of silver. The duality of myth and metaphor gives rise to a contradiction: "It happened/it did not happen (thus), it is/it is not (thus)." This is precisely the same duality we found in the paradoxes of Epimenides, Quine, and Groucho Marx mentioned above. And since recursivity gave birth to language, whenever language returns to its origins, it retraces its path scattering the way with a string of paradoxes.

EPILOGUE: WHERE IS REALITY?

This chapter has been an invitation to wander among the frames of human existence generated by recursivity, an itinerary that could be said to evoke that of "Pictures at an Exhibition" by the Russian composer Mussorgsky: a brief stroll to the measure of a motif punctuated by moments of contemplative pause. But this fascinating panorama gives rise to a colossal enigma: Recursivity surrounds us facilitating links and connections, but where is reality? Does reality really exist? If, as suggested by Escher's "Print Gallery," human existence is a kaleidoscope (a succession of frames) in which there is an indisputable place for all the forms of imagination, creativity and art, such as play, theater, myths, rituals, and narrations (the novel), that have their roots in the distinction between the map and the territory and in all of which fantasy reigns, we cannot be sure that reality—what folk psychology calls "reality"—is any more real than what we experience through those scenarios. Perhaps the only memorable reality is that of the human mind capable of living, recursively, so many and such different realities.

NOTES

1. A "porrón" is a glass vessel with an elongated conical spout similar to that of a teapot on one side. Drinkers position the spout close to their mouths (without touching it) and tip the glass jar, catching the liquid in their mouths as it emerges from the spout. Its use is traditional in Spanish taverns, allowing several drinkers to share the jar and making glasses unnecessary.
2. Gödel's theorem, which has broad mathematical and philosophical implications, can be formulated in accessible language as follows: within any formal system (mathematics), there are correctly formulated expressions that cannot be proved true or false. To get an idea of what the theorem implies, consider the well-formed expression $x^2 + 1 = 0$, which in the algebra of real numbers has no solution (it is meaningless). Only when the domain of the numbers is enlarged to include the imaginary unit i will there be a solution, but this solution will be in the domain of complex numbers. This enlargement of a system by virtue of which an absurd numerical expression has a solution is called Gödelization.

3. This is a good place to introduce the mathematical expression for recursivity: It is the reiterated application of a function $f(x)$ to itself : $f(f(f(x)))$ etc. Note that a confusion often arises, even in the academic literature, between recursivity strictu sensu and the linear repetition of a process. We express the latter "reiteration" graphically as: $f(x)$ & $f(x)$ & $f(x)$, while the recursive "reiteration" is a function: $f(f(f(x)))$.
4. Certain pathological cases, by contrast, illustrate the impossibility of dealing with two planes and, therefore, of becoming an observer: The delirium of a mother who, having lost her child, cuddles and rocks a doll in her arms. Her fantasy is her reality. She enters into a frame and is trapped inside it. She cannot escape. (I owe this interesting insight to Marta Sadurní).
5. One version of Quine's paradox is the statement, "This sentence in quotes is false" or a footnote that says "Do not pay any attention to the footnotes." The example most often encountered in daily life is "Don't pay any attention to what I tell you to do."
6. In a role-playing game, a participant may lose the notion of "this is play," and the situation can degenerate dramatically. Children engaged in a play fight can end up really hitting and hurting each other.

REFERENCES

Bateson, G. (1972). A theory of play and fantasy. In *Steps to an ecology of mind.* New York, NY: Ballantine Books. (Original work published 1955)

Donald, M. (1991). *Origins of the modern mind.* Cambridge, MA: Harvard University Press.

Frye, N. (1990). *Words and power.* New York, NY: Harcourt, Brace and Jovanovich.

Hofstadter, D. (1979). *Gödel, Escher, Bach: An eternal, golden braid.* New York, NY: Basic Books.

Lalueza, J. L. (1991). *Desarrollo del símbolo en el juego interactivo en niños con Síndrome de Down y sin disminución.* Unpublished doctoral theses, Universidad Autónoma de Barcelona.

Lalueza, J. L., & Perinat, A. (1994). Desarrollo de los significados compartidos en el juego entre los adultos y los niños con Síndrome de Down. *Infancia y Aprendizaje, 67–68,* 133–146.

Leslie, A. (1987). Pretense and representation: The origin of theory of mind. *Psychological Review, 94,* 412–426.

Maturana, H., & Varela, F. (1980). *Autopoiesis and cognition.* Boston, MA: Reidel.

Maturana, H., & Varela, F. (1990). *El árbol del conocimiento.* Madrid, Spain: Debate.

Perinat, A., & Sadurní, M. (1995). "Juguemos a llamar por teléfono." Juego simbólico y procesos recursivos en la interacción comunicativa. *Substratum, 3*(7), 77–102.

Perinat, A., & Sadurní, M. (1999). The ontogenesis of meaning. An interactional approach. *Mind, Culture and Activity, 6,* 53–76.

Rumbaugh, D. M. (1977). *Language learning by a chimpanzee. The LANA Project.* New York, NY: Academic Press.

Sadurní, M. (1992). *La ontogénesis del significado.* Unpublished doctoral thesis, Universidad Autónoma de Barcelona.

Sadurní, M. & Perinat, A. (1994). El proceso ontogenético de la significación. *Substratum, 2*(5), 17–39.

Savage-Rumbaug, S. (1986). *Ape language : From conditioned response to symbol.* New York, NY: Columbia University Press.

Savage-Rumbaugh, S., Murphy, J., Sevcic, R. A., Brakke, K. E., Williams, S. L., & Rumbaugh, D.M. (1993). Language comprehension in ape and child. *Monographs of the SRCD, 58*(3–4).

Varela, F. (1989). *Autonomie et connaissance.* Paris, France: Le Seuil.

CHAPTER 2

TURTLES ALL THE WAY DOWN?

Recursion and Infinity in the Human Sciences

Lee Rudolph

This chapter began (and begins) as a survey, from one mathematician's point of view, of how of the terms *recursion*, *recursive*, and the like are used in mathematics and in the human sciences. That survey, and an analysis of its results, led (and leads) to a similar survey and analysis of the various uses of *infinity*, *infinite*, and the like.

The first two sections of the chapter are mostly about mathematics as a social and cultural activity. I give some history of mathematicians' use of the terms *recursion* and *recursive*. I discuss differences of form and function among ellipsis and aposiopesis in general discourse, in typical "scholarly"/"scientific" discourse, and in "paramathematical" discourse. I draw various connections among aposiopesis, infinity, well-foundedness, recursion, and computation in mathematicians' discourse and (other) behavior. I describe the history and nature of the "horror of infinity" and its

Cultural Psychology of Recursive Processes, pages 23–101

not always recognized twin, enthusiasm for (or at least pleased acceptance of) infinity. I conclude with speculation about the role that two axioms first introduced during the mathematical formalization of set theory between 1874 and 1925, the Axiom of Infinity and the Axiom of Foundation, might have played in essentially eliminating the "horror of infinity" among mathematicians.

The last two sections of the chapter are mostly about the human sciences, and human experience more generally. I describe and distinguish several families of uses (not too closely related to each other) of the terms *recursion* and the like in the human sciences and trace their lineages back to "base cases" (not all of them in mathematics), then do the same for *infinity* and the like. I observe that in the human sciences there is considerable enthusiasm for infinity, although "horror of infinity" persists here and there; I argue that the enthusiasm is misplaced and that the horror is unnecessary. The former argument—partly empirical, partly theoretical—arises out of a discussion of finities and infinities in human experience, framed in terms of some axioms of "evolutionary ontology," including von Uexküll's Axiom of Subjective Finiteness. I only sketch the latter argument; it applies Aczel's nonstandard set-theoretical Anti-Foundation Axiom to derive a formalism for infinity-free modeling of several lineages of 'recursion' in the human sciences.

RECURSION IN MATHEMATICS

The history of "recursion" in mathematics is deeply entangled with that of computation, as well as with the development of formal, fully mathematicized notions of infinity, both actual and potential. In this section I make an effort at disentanglement.

The Increasing Formalization of Recursion in Mathematics

To explicate the usual notion of "recursion" in modern mathematics,[1] I must first say something about the notion of "function" in modern mathematics. There are various mathematically equivalent (though superficially very different) formal definitions of a function. Informally, they all come down to the following notion: A function f is defined by a rule that assigns to certain values of x (the allowable *inputs* of f; its *independent variables*) corresponding values $f(x)$ (the realized *outputs* of f; its *dependent variables*).

It is of the utmost importance to realize that here "rule" has to be understood extremely generally. A rule could be, for instance:

- a formula (e.g., the algebraic formula x^2 defines a function f for which an allowable input x is something that can be squared—typically a number—and the output $f(x)$ is the square of x)
- a "look-up table" (e.g., a telephone directory defines a function f for which an allowable input x is the name of a subscriber listed in the directory, and the output $f(x)$ is the list of telephone numbers listed for subscriber x)
- a computer program (e.g., an income-tax calculator defines a function f for which an allowable input x is the totality of data the user must supply to comply with the requirements for filing a given tax form, and the output $f(x)$ is the totality of obligations placed on the user when the tax code is applied to that data x)

The one nonnegotiable feature that a rule must have, to qualify it to define a function f, is that if x_1 and x_2 are allowable inputs that are equal to each other (despite having different names, and perhaps being accessed in different circumstances, e.g., at different times), then the outputs $f(x_1)$ and $f(x_2)$ must also be equal; otherwise put, to define a function, a rule must be reliable (non-aleatoric, deterministic, unambiguous, etc.).

Note that a computer program can include formulas and look-up tables (an income-tax calculator certainly includes many of both), and a single rule may have a complex internal structure comprising many intricately (but deterministically) linked subrules. Note also that the correspondence between rules and functions is not one-to-one, it is highly many-to-one: given a function, there are *always* an unlimited number of different rules that define it, no two of which bear any obvious relationship to each other (that two rules do in fact define the same function is rarely obvious, and in some cases can only be proven with extreme effort,[2] if at all).

Now I can give an example of a function F "defined by recursion." The allowable inputs of F are the "natural numbers" 1, 2, 3, ... The following multicase rule defines F by giving a definite procedure to calculate the output value $F(n)$ for each natural number input value n.

$$F(1) = 1;\ F(2) = 1;\ \text{if } n > 2 \text{ then } F(n) = F(n-1) + F(n-2).$$

As an illustration, here are the values of $F(n)$ for n running from 1 to 5, and how they are calculated, using the given procedure.

$$\begin{aligned} F(1) &= 1,\ F(2) = 1 \text{ (stipulation)} \\ F(3) &= F(3-1) + F(3-2) = 1 + 1 = 2 \\ F(4) &= F(4-1) + F(4-2) = 2 + 1 = 3 \\ F(5) &= F(5-1) + F(5-2) = 3 + 2 = 5 \end{aligned}$$

Evidently the values $F(n)$ for all n are determined by this procedure in precisely the same manner (at least, in an ideal world where time, space, memory, and other resources are unlimited), so the given rule does, in fact, define F as a function. This is a prototypical "definition by recursion."

$F(n)$ is called the nth Fibonacci number; F may be called the Fibonacci function. Algebraic manipulations show that for every n, $F(n)$ is the natural number obtained by rounding down the 'real algebraic number'

$$(2\times\varphi^n+\sqrt{5})/(2\times\sqrt{5})$$

(where φ is the real algebraic number $(1+\sqrt{5})/2$, often called the Golden Ratio) to a natural number. This is a prototypical "definition by formula."

The qualitative distinction between the two definitions of the Fibonacci function is very clear. The definition by recursion stipulates:

R1 a list of *base cases*—as it were, a look-up table to be used to determine what output $F(n)$ the function F is to assign to each input n specified in that list; and

R2 a *reduction rule* to be used to determine what output $F(n)$ the function F is to assign to each input n not specified in that list.

What makes the rule specifically a *reduction* rule (and the definition a definition *by recursion*) is that:

R3 the inputs of F belong to a *hierarchy of levels* that is

(a) *transitive* in the sense that if L, M, and N are levels of the hierarchy such that M is subordinate to L and N is subordinate to M, then N is subordinate to L,

(b) *well-founded* in the sense that no series of different levels L_1, $L_2, \ldots, L_k, \ldots$, with L_k subordinate to L_{k-1} for every $k > 1$ can go on forever: every such series must terminate at a finite stage with a base level (one to which no other level is subordinate), and

(c) *based* in the sense that each input specified in any of the listed base cases belongs to a base level, and each input that belongs to any of the base levels is specified in a listed base case; and

R4 the rule that determines the output $F(n)$ for a given input n that is not given in the list of base cases must make use only of outputs $F(k)$ for which the level of the input k is strictly subordinate to the level of the input n—that is, the rule relies on *reducing* the hierarchical level.

Note that such a "hierarchy of levels" need *not* be linear: It is perfectly possible (and very often the case in practice) that some two levels L and M are incomparable: neither L nor M is subordinate to the other.

Conditions **R1–R4** are easily checked for the definition by recursion of the Fibonacci function:

R1 The list of base cases comprises two cases, $F(1) = 1$ and $F(2) = 1$.
R2 The reduction rule is the "two-term linear recurrence relation" $F(n) = F(n-1) + F(n-2)$.
R3 The inputs of F are the natural numbers $n = 1, 2, \ldots$, arranged in the usual (linear) hierarchy, where, as levels, p is subordinate to q if and only if $p < q$ as numbers—which is obviously (a) transitive and (b) well-founded—and (c) the base levels are 1 and 2.
R4 The reduction rule reduces the hierarchical level (by 1 and by 2).

In contrast, at least on its face[3] the definition of the Fibonacci function by formula uses no base case (nothing is looked up), and its rule (that is, its formula) neither performs a reduction nor requires a hierarchy: by mere algebra, a given value $F(n)$ can be computed with no reference to, or knowledge of, any other values $F(\mathrm{k})$ with $k \neq \mathrm{n}$. The definition by formula seems more direct, somehow, but at the same time less "hands-on."

The scheme laid out in **R1–R4** was, in essence, introduced to modern mathematics by Dedekind (1888), who called it "*Definition durch Induction*" ("definition by induction") in the body of his text (p. 35, ¶ 126) and glossed it "*Definition durch Induction (oder Recursion)*"—"definition by induction (or recursion)"—in the preface to the first edition (p. 6, in a forward reference to ¶ 126).

Of course, as Soare (1996, p. 287) points out, "Well before the nineteenth century mathematicians used the principle of defining a function by induction." In fact, an exposition of the principle of "mathematical induction" (a phrase first used by De Morgan, 1838, p. 466, though his preferred term—never generally adopted—was "successive induction") has been found by Rabinovitch (1970) in a Hebrew textbook on arithmetic calculation from the early 14th century C.E. (Levi ben Gershon, 1321/1909); there it was called *hadragah* (הדרגה), literally "rising step by step."

Thus, Dedekind's contribution in laying out such a scheme as **R1–R4** was less to innovate than to formalize and unify two previously more or less informal, more or less distinct mathematical practices.

Soare (1996, p. 287) summarizes the situation after Dedekind (1888): "Based on this work, Peano" (1889, 1891) "wrote the familiar five axioms for the positive integers [i.e., natural numbers]. As a companion to his fifth axiom, mathematical induction, Peano used definition by induction, which has been called primitive recursion" since 1934.

Gödel (1934) gave a formal definition of what later came to be called (general) recursive functions, based on Herbrand (1931/1971); it was immediately recognized as the "right definition" by everyone then working in the field, and remains the basis of recursion theory today (an excellent example of the "benign fixation of mathematical practice" named and described by Azzouni, 2006, and further discussed by Rudolph, 2012a), which is itself the basis of theory of computation, pioneered by Turing (1936).

Inclusion of "Termination" in Idealized Computations

Certainly, on the available evidence, as with all the rest of their activities, the computations and calculations of non-ideal humans eventually do terminate. That does not make it obvious that a mathematical definition of "computation" must include the stipulation that every computation must terminate. On the other hand, as just seen, termination is built into the mathematical definition of "recursion" by the well-foundedness assumption **R3**(b). Thus the inclusion of "termination" in a mathematical definition of "computation" might be justified, without appeal to mortality of (even idealized) human calculating agents, if it were so that computation (and what can be done with it, i.e., what is "computable") and recursion (and what can be defined using it, i.e., what is "recursive") have exactly the same meaning—or as exactly the same as is possible, given that "computable" already has an informal meaning (even in mathematics), whereas "recursive" has in present day mathematics (precisely: since 1934) a formal meaning only.

Soare (1996) has given a definitive scholarly survey of the historical and logical connections between "computation" and "recursion" (to the end of bolstering his case for four recommendations on their use and related issues).[4] Soare sets his terms in his survey's abstract.

> We consider the informal concept of "computability" or "effective calculability" and two of the formalisms commonly used to define it, "(Turing) computability" and "(general) recursiveness", [. . . W]e make several recommendations [. . .] about preserving the *intensional* differences between the concepts of "computability" and "recursion." (p. 284, italics in the original)

Eighteen pages of truly fascinating scholarly detail later, Soare states his first recommendation, prefaced by a brief summary of current usage—from which I have cut some irrelevant technicalities—as established in those pages (italics in the original).

> The term and concept of "computable" is associated with the notion of computation (§2.1), algorithm (§2.3), and with the functions defined by (or sets enumerated by) Turing machines (§3.1) or register machines (§3.4) [etc.].

> The term and concept of "recursive" is associated with: definition by recursion (induction) (§2.2), general recursive functions in the sense [of] Herbrand-Gödel (§2.4) [etc.]
>
> RECOMMENDATION 1. *The term "recursive" should no longer carry the additional meaning of "computable" or "decidable."* (p. 312)

What Soare here glosses as "the notion of computation (§2.1)" begins with a semi-formal consensus definition.

> A *computation* is a process whereby we proceed from initially given objects, called *inputs*, according to a fixed set of rules, called a *program*, *procedure*, or *algorithm*, through a series of *steps* and arrive at the end of these steps with a final result, called the *output*. The algorithm, as a set of rules proceeding from inputs to output, must be precise and definite, with each successive step clearly determined. (Soare, 1996, p. 286; italics are in the original and indicate definitions, not emphasis)

Soare's abstract emphasizes "*intensional* differences," and his recommendation calls for taking action to preserve those differences, precisely because *extensionally* there is no difference between the two "formalisms commonly used to define" (that is, to formalize) "the informal concept of 'computability'"—namely "(Turing) computability" and "(general) recursiveness": although the formalizations are entirely different, their ranges of application are provably identical.

This identity of extension follows from three propositions: (1) Something is "intuitively computable" if and only if it results from some computation in the sense of the semi-formal consensus definition, which stipulates termination: a computation must "arrive at the end [...] with a final result." (2) "The Church-Turing Thesis that the intuitively computable functions coincide with the formally computable ones" (Soare, 1996, p. 285) is true. (3) The (formal, universally accepted) definition of "formally computable" functions, namely, "recursive functions," does stipulate termination (suitably formalized).

Eppur non si finita

Gandy (1988, p. 67), when he writes "The use of recursion in computation is at least as old as Euclid," certainly refers to the 'Euclidean algorithm' that is laid forth in the proofs (but not the statements) of Propositions 1 and 2 in Euclid's *Elements*, Book V (Heath, 1908, vol. 1, pp. 296–300). Without going into the mathematical details of the Euclidean algorithm (annotated, animated versions of the proofs have been published by Joyce, 1998), I use the algorithm to question whether it is quite fair for Gandy (1988) to use it as an example of "computation" in his sense (which is the sense later endorsed by Soare, 1996, and quoted above).

It is not in dispute that the Euclidean algorithm *can* "arrive at the end [...] with a final result," given suitable inputs. Here "suitable" encompasses both certain inputs of a geometric nature (like those that Euclid envisions in his statements of Propositions V.1 and V.2, namely, an arbitrary couple of "commensurable magnitudes") but also certain inputs of an arithmetic or algebraic nature, like an arbitrary couple of natural numbers m and n. In modern practice, such arithmetic or algebraic inputs are those to which the Euclidean algorithm is essentially always applied, often under the name of "greatest common divisor algorithm."

The algorithm is readily reformulated as a process to calculate the value $G(x, y)$ of a function G of two numerical variables x and y in terms of base cases (to wit, $G(x, x) = x$ for all x) and reduction rules (to wit, $G(x, y) = G(y, x)$ for all x and y, and $G(x, y) = G(x - y, y)$ for all x and y with $x > y$). So formulated, however, the process need not terminate without further restrictions on x and y, the reason being that, without some such restriction (that x and y be natural numbers is sufficient, but far from necessary), the transitive, based hierarchy of inputs *need not be well-founded.*

In fact, many interesting cases in which the process *does not* terminate were well known to Euclid, perhaps the simplest being that with x equal to 1 and y equal to the square root of 2. Proposition 2 of Book X of the *Elements* (Heath, 1908, vol. 3, p. 17), a "test for incommensurable magnitudes" (of which 1 and the square root of 2 were the first pair known to Greek mathematics), is summarized by Heath in his commentary as follows: "The sign of the incommensurability of two magnitudes is that *this operation never comes to an end,* while the successive remainders become smaller and smaller until they are less than any assigned magnitude" (1908, p. 18; italics in original).

Regardless of whether a couple of inputs x and y (positive real numbers) are both natural numbers, the "set of rules" is "precise and definite, with each successive step clearly determined." The question arises, why—other than convention (subsumed into "the benign fixation of mathematical practice")—should the term "computation" *not* be used for a process "whereby we proceed from initially given [...] inputs, according to a fixed set of rules, [...] through a series of steps," *even if*—as when the Euclidean algorithm is given incommensurable inputs—we never "arrive at the end of these steps"?

It is noteworthy that nowhere does Turing (1936)—the ultimate source to which Soare (1996, pp. 291–294) refers the essence of the semiformal consensus definition of a computation quoted above—make any mention of a computation coming to an "end" (or anything similar, e.g., "stopping," "terminating," "reaching a final state," etc.). On the contrary, Turing writes "[t]he 'computable' numbers may be described briefly as the real numbers whose expressions as a decimal are calculable by finite means," and again "[a]ccording to my definition, a number is computable if its decimal can be written down by a machine" (1936, p. 230), where it is clear that by "decimal" Turing intends "not necessarily terminating decimal expansion."

Since Turing's "computable" numbers do include numbers (like the square root of 2) with provably *non*terminating decimal expansions, it appears to me that for Turing a "computation" might (and typically would) continue forever, even though his "machine" (what he formalizes on p. 232 of his paper as an "automatic machine," first called a "Turing machine" by Church, 1937) is finite and calculates "by finite means."

If talk of "machines" in the context of the human sciences seems out of place, note that Turing (1936) actually introduces his "automatic machine" as a formalization (thoroughly mathematical, though described in suggestive mechanistic terms like "tape" and "scanning") of "an idealized *human* calculating agent" (Soare, 1996, p. 291, italics in the original), called by Turing a "computer." Why (except, indeed, because the associated *mathematical formalization* has powerful and convenient consequences) should "an idealized *human* calculating agent" be stipulated to have performed a computation only when the task at hand has been completed?

Surely it is more "*human*"—and not *a priori* less "ideal"—to allow that at least *some* "computations" do *not* terminate. As Turing remarks, "It is always possible for the computer to break off from his work, to go away and forget all about it, and later to come back and go on with it" (1936, p. 253). It seems to me that then it must also be "always possible for the computer to break off" and *never* "come back" (in fact, this often happens in the lives, and invariably upon the deaths, of non-idealized human calculating agents). Equally, it seems to me that it must be "always possible for the computer" to "go on with it" *indefinitely without termination* (this too has been known to happen in human life), and so—by idealization—to "go on with it" *forever.* If so, what should we call the "it" that is being "go[ne] on with" forever, if not a computation?

My comments in this subsection are not intended to challenge the inclusion of the stipulation that every computation must come to an end in the semiformal consensus definition. Such a challenge would be more than merely quixotic—it would be tilting at a straw windmill, for as Gandy points out (1988, p. 82, citing Kleene, 1981, p. 60), already by 1936 "it had been realized" (by Kleene and others in the Princeton circle also including Turing, Church, and Gödel) "that one had to deal with computations which might not terminate," a realization that "did not get into print until" Kleene (1938).

So far as I can tell (as a mathematician who is not a recursion theorist), recursion theorists finesse this point by leaving it understood (without stating it in the semiformal consensus definition) that the "initially given objects, called *inputs*" of a computation must be restricted to (at most) those "given objects" for which a stop condition of one of the "fixed set of rules" will be reached at some step of the process whereby we proceed from those given objects.[5] In other words, it is an accepted presupposition of the definition that the inputs of a computation belong to a level hierarchy that is (not only transitive and based, but also) well-founded.

ELLIPSIS, APOSIOPESIS, AND INFINITY

Infinity has already sneaked into this chapter, hiding behind "..." in the printed phrase "the 'natural numbers' 1, 2, 3,..." (above the recursive definition of the Fibonacci function, p. 25). Typographical custom, when it calls "..." "ellipsis points," thereby conflates two distinct functions of "...": to indicate *ellipsis* and to indicate *aposiopesis*. In this section I explicate the differences between those two functions in various discourse communities.

Ellipsis and Aposiopesis in General and Scholarly Discourse

In general speech and writing, ellipsis is the "omission of one or more words in a sentence, which would be needed to complete the grammatical construction or fully to express the sense" (ellipsis, 1891/2013). Its rhetorical function is to invite or coerce the listener or reader to complete the expression of the sentence's "sense"—whether to (somewhat) relieve the speaker or writer of responsibility for doing so (at that point), to introduce deliberate ambiguity (which may or may not be resolved at a later point), or as an aid to some other rhetorical task involved in the narrative negotiation.

Aposiopesis is a "rhetorical artifice, in which the speaker [or writer] comes to a sudden halt, as if unable or unwilling to proceed" (aposiopesis, 1885/2013). Its function is to convey to the listener or reader the speaker's or writer's true or feigned inability or unwillingness to proceed, whereby it can produce further effects, including "powerful affect-channeling" effects (Burke, 2010, p. 130), which I ignore here. Aposiopesis need not occur at the end of a sentence, though it usually does; in printed prose, not all ellipsis points at the end of a sentence indicate aposiopesis, though most do.

In general speech, including general writing read aloud, ellipsis and aposiopesis are usually performed using (different) tricks of the voice, etc., rather than words (which, in reading aloud, would have to be interpolated; a commonly favored alternative is to render ellipsis points as "dot dot dot").

Discourse analysts widely agree that various presuppositions (e.g., to a first approximation, the "maxims" of "Quality, Quantity, Relation, and

Manner" enunciated by Grice, 1989, pp. 30ff.) are fundamental to most discourse communities and genres, but also that such presuppositions are typically malleable, subject to being overridden in the pursuit of certain discourse functions. Thus, for instance, Grice (1989, p. 30) distinguishes what it means to "flout" one of his maxims ("blatantly fail to fulfill" it) from what it means to "violate" one (to break it "quietly and unostentatiously"). When a speaker or writer flouts a maxim, the reader or listener can be expected—precisely because of the blatancy of the failure to fulfill it—to notice and interpret the failure (in expected, or possibly unexpected, ways); violating a maxim is more "liable to mislead," often deliberately.

Although discourse communities (more or less by definition) deprecate both the flouting and the violation of their (defining?) presuppositions, both flouting and violation of presuppositions are (more or less universally) available resources for discourse community members, though of course subject to sanctions—those "(sometimes brutal) social tools to induce conformity that routinely appear among us *whenever* behavior really is socially constrained" that Azzouni (2006, p. 208; italics in the original) finds oddly absent from "mathematics as a social practice." In (all?) discourse communities, the "social tools to induce conformity" to such presuppositions cover a wide range, from mild rebuke up to expulsion or worse.

In contemporary scholarly writing in general, and perhaps especially scientific writing, the presuppositions include:

- originality of language and ideas (plagiarism is forbidden)
- honest attribution (citations of others' or one's own work must be complete; quotations must be correct and complete, with only truly irrelevant matter omitted)
- no deception (data must be neither false nor misleadingly presented; matter omitted from quotations *must be replaced with ellipsis points*)

I have the impression that, in these writing communities, ellipsis points are essentially never used for any purpose other than to comply with this last presupposition. If they appear at all in such writing, rhetorical ellipsis and aposiopesis are indicated by using words, not punctuation marks.

The situation is entirely different as regards *paramathematical* writing—that is, writing on mathematical subjects, which includes but is not limited to fully mathematical writing written by mathematicians for mathematicians.[6]

Of course, ellipsis points can be used within quotations in paramathematical writing just as in general scholarly writing (although, as a matter of fact, paramathematical writing rather rarely includes quotations, and its subset of fully mathematical writing hardly ever does). But paramathematical writing carries an additional, very important presupposition: Ellipsis points can also be used to indicate "an omission of one or more words in a sentence"—or, importantly, in a formula—if, *and only if,* the intended reader has both the necessary information (from implicit or explicit context) and the necessary ability "fully to express the sense" correctly (this can be read as a very strong codification of several of Grice's maxims).

I propose that the rhetorical function of ellipsis points in paramathematical writing is neither to invite nor to coerce, but to remind the reader of this presupposition and of the corollary responsibility *to use the necessary information and ability* (not necessarily immediately, and perhaps only in—well-grounded—imagination, cf. pp. 63–72). Not only must "[e]verything in a publication [. . .] be based on the individual understanding of the author, nothing being accepted on authority, no matter how distinguished" (Bonsall, 1982, p. 9), but (by a natural, and usual, extension of Bonsall's idealization of the community of mathematicians) *everything read in a publication must* (eventually, and in any case before being used by the reader in public or in print) *be understood by the reader,* "nothing being accepted on authority," *even on the writer's authority.*

Just as in other writing, in paramathematical writing ellipsis points can occur either in the middle of a sentence (or formula) or at its end (more rarely, its beginning). In paramathematical writing, there is a big difference between these two uses; it can be detected by observing mathematicians' behavior, and explained (I claim) using the semi-formal definition of *computation,* taken semi-literally—neither utterly algorithmically (in the style of "hard computer science"; cf. Pascual-Leone, 1997), nor (merely? mostly?) metaphorically (which seems to be the style of at least some "soft" cognitive science; cf. Rudolph, 2006a, 2006b).

Paramathematical Ellipsis Points, Computation, and So On . . .

When mathematicians read paramathematical medial ellipsis points aloud, they invariably (in my experience) use one of a small store of set phrases, rather than some highly context-dependent phrase. Here are a few examples.

- Outside some extraordinary context, the written expressions "1, 2, . . . , 10" and "1, 2, 3, 4, 5, 6, 7, 8, 9, 10" have the same clear (extensional) meaning. But when "1, 2, . . . , 10" is read aloud (by a mathematician), it does not become "one, two, three, four, five, six, seven, eight, nine, ten"; rather, it becomes *[from]* one *[up] to* ten or the like (the small variations apparently depending on syntactic and pragmatic/considerations, not semantic ones). Similarly, when

"1, 2, . . . , 1000000" and "1, 2, . . . , $10^{10^{100}}$" are read aloud, they do not become complete and uninterrupted recitations of the series of names of the first one million natural numbers (which is humanly improbable) nor of the first googolplex (which is physically impossible); they become something like *from* one *to* ten and *from* one *to [a]* googolplex.

- When the context establishes (for the intended reader) that m and n are natural numbers with $m \leq n$, the written expressions "$m, \ldots, n$" and "the natural numbers from m to n" are synonymous. Similarly, the written expressions "$2^m, \ldots, 2^n$" and "the powers of 2 from the mth to the nth" are synonymous. But when "$m, \ldots, n$" and "$2^m, \ldots, 2^n$" are read aloud (at least by this mathematician; nor do I recall having observed other mathematicians behave otherwise) they do not become recitations of anything like the respective paraphrases just given—rather, they become something like *from* m *to* n or *from* 2^m *to* 2^n.

It is my impression that, as in these examples, essentially all medial ellipsis points in paramathematical writing occur in expressions referring to sets with *finitely many members.* I do not think it is far-fetched to interpret these examples as evidence that the function of medial ellipsis points in paramathematical writing is not to *abbreviate* (some possible notation for) *a finite set* (namely, the extension of the written expression within which they are a constituent); it is to *stand in for a process* that the reader (by presupposition) has the knowledge and ability to reconstruct and that would produce the required finite set in a finite number of steps: in short, a terminating computation.

What about terminal (or, rarely, initial) ellipsis points in paramathematical writing? In my experience, when mathematicians read these aloud, they invariably use one of a small store of set phrases that are *entirely different* from those used for paramathematical medial ellipsis points. Although other examples exist in moderate profusion, the prototypical paramathematical terminal ellipsis points are those in the written expression "1, 2, 3, . . . ," and I limit my further discussion to that case.

When the written expression "1, 2, 3, . . ." is read aloud, the ellipsis points are usually rendered by the words "and so on" or "and so forth" (less often as "et cetera," and very informally as "dot dot dot"). What does that mean?

Grammatically, *so* is an "adverb of process" (so called by descriptive linguists; cf. Quirk, Greenbaum, Leech, & Svartvik, 1985, p. 865) and the set phrase *and so on* is a "vagueness marker" (so called by discourse-analytic functional linguists):

> There are only four phrases included in the AFL [Academic Formulas List] that are classified as vagueness markers, making it the smallest functional category. Furthermore, three of these phrases are limited to the Spoken AFL; only the phrase *and so on* appears in the Core AFL. Nevertheless [...] making vague references with these particular extenders is a common discourse function in academic speech. (Simpson-Vlach & Ellis, 2010, p. 505, italics in the original)

The question becomes: In "1, 2, 3, . . . ," what is the "process" and what function does its "making vague references" perform (and for whom)?

The process is what Whitrow (1933) calls "the law of ordinal arithmetic, viz. that every integral [i.e., natural] number has a successor." This "law" has multiple aspects, at least two of which Whitrow certainly had in mind.

(1) For a modern mathematician (like Whitrow, in 1933 a 21-year-old Oxford graduate with a fresh "double-first" in mathematics), "every integral number has a successor" is an informal statement of one of the formal "familiar five axioms" (Soare, 1996, p. 287) universally called the "Peano postulates" (at least in English-language mathematics, starting with Huntington, 1905) that were elaborated in various formulations during the second half of the 19th century by Grassmann (1861), Dedekind (1888), and Peano (1889, 1891).

(2) For a historian and philosopher of science (like Whitrow; cf. Whitrow, 1932), "every integral number has a successor" recapitulates a fact of the development of "ordinal arithmetic" both as a practice and as a (proto-)mathematical theory—this last notably among the classical Greek philosophers.

Under aspect (1), the reference is not "vague" so much as it is informal: Its intended formal meaning is as precise as any piece of formal mathematics can be. At an intermediate level of formality, using English words and a minimum of mathematical symbols (in place of the "Latine sine flexione" and idiosyncratic glyphs favored by Peano, 1899), and incorporating all five Peano postulates, the intended meaning of "1, 2, 3, . . ." is

> the one and only set **N** (up to structure-preserving isomorphism) that is characterized by (I) containing a special element 1, and (II) containing for each of its elements *i* an element $s(i)$, where *s* is a function from **N** to **N** with these properties: (III) for any elements *i* and *j* of **N**, $s(i)$ equals $s(i)$ if and only if *i* equals *j*; (IV) there is no element *i* of **N** for which $s(i)$ equals 1; (V) if *K* is a set of which each element is an element of **N**, and 1 is a member of *K*, and for every *i*, if *i* is a member of *K* then $s(i)$ is a member of *K*, then *K* equals **N**.

Thus, under aspect (1) the function of the ellipsis points in "1, 2, 3, . . ." for a paramathematical writer is to put the intended reader in mind, not just of the static extension **N** of that written expression, but of the dynamic, non-terminating process—the prototypical non-terminating computation upon which (in quite a precise sense) all others are built—that generates it. I do not think it is inappropriate to describe this as aposiopesis in the limited sense of "a sudden halt, as if unable [. . .] to proceed" (aposiopesis, 1885/2013) *all the way to completion*, where the inability is literal, and "willing"ness is irrelevant.

Under aspect (2), the reference of *and so on*, unextended by qualifiers, is maximally vague: it makes no commitments, and thus need honor none. The phrase has several common extensions, chiefly *and so on* [or *forth*] *forever, and so on to infinity,* and *and so on* [or *forth*] *indefinitely*[7]; these are less than maximally vague, in different ways. The extension by *forever* puts the writer (or speaker) in the position of looking into an endless future that is vaguely delimited except for an unavoidable commitment that the process of counting *will* go on, "step by step." The extension by *to infinity* may be vaguer (in that what is endless is not explicitly assigned the temporal quality of duration) or merely different (since, arguably, what is endless is implicitly assigned the spatial quality of extent). The extension by *indefinitely* seems (to me) to be without either temporal or spatial qualities; it suggests, rather, mild chaos and a lack of firm rules.

Thus, under aspect (2) the ellipsis points in "1, 2, 3, . . ." need not put the paramathematical writer's reader in mind of a non-terminating computation, but may very well summon up for that reader images of unbounded space or time, chaos, and infinity. None of these images has intrinsic affect, but the historical record makes it clear that affect—both positive and negative—can attach to them. I do not think it is inappropriate to describe this, too, as aposiopesis, where the inability to proceed (to completion) is still literal, but "willing"ness is now in question as well.

Odi et Amo

The case of negative affect has had more press. Indeed, it even has a Latin name, *horror infiniti*—though it had already been felt by some Greek philosophers, e.g., Zeno:

> I think that to him the source of the dilemma in these paradoxes appeared to be the law of ordinal arithmetic, viz. that every integral number has a successor, for was it not this which made possible the existence of an infinite sequence? In him it was not "horror of the infinitesimal and indivisible" but "horror of the infinite process" which the Eleatic arguments inspired. (Whitrow, 1933, p. 153)

Yet Aristotle (who did use the phrase "καὶ τοῦτ' εἰς ἄπειρον," Aristotle, 1902, p. 73, translated literally as "and so on to infinity" by Owen, 1853, p. 274) takes "infinity" in his stride, disapproving in many contexts but approving in some. According to Mendell (2004), Aristotle distinguishes *potential* infinities from *actual* (or *completed*) infinities, accepts potential infinities "by division [...] as central to his notion of continuous magnitudes," and both potential and actual infinities "by addition" *as applied to time* ("there is always another day"), and rejects both potential and actual infinities "by addition" *as applied to space, to geometrical entities* (lines, planes, solids), *or to material objects* (at least, in this case, when they lead "to a magnitude greater than the magnitude of the universe," which is "finite in size and weight") and actual infinities "by division" as applied to geometric entities.

These two views of infinity—distrust sliding into "horror" (especially of actual infinities), and acceptance (especially of potential infinities) sometimes verging on enthusiasm—persist to the present day: certainly in general discourse (including books, magazine articles, radio and television programs, and websites devoted to popularizing science), but also in some scholarly discourse, particularly in the human sciences. On the other hand, full-blown horror of infinity has completely disappeared from mathematical (and most scientific) discourse, and at least in mathematical discourse any remaining distrust of infinity (e.g., Nelson, 1995, 2011) has been subsumed into "the benign fixation of mathematical practice"—mathematicians can (if they choose) see what they can do by (for instance) eschewing actual infinities, but what they do is only "developing more mathematics, not [...] changing [its general] practice" (Azzouni, 2006, p. 206).

It seems to me that the diminution in "horror of infinity" (equivalently, the increased acceptance of infinity, either potential or actual) springs from different sources according to the community in which it has taken place.

In the general intellectual community, and among scientists (including at least some human scientists), the source seems to me to have been part of a larger shift in general culture, the mechanistic universe of Deism coming to terms with (if not being overwhelmed by) the universe of natural selection and laissez-faire. What is a "divine watchmaker" other than an idealized calculating agent (albeit not human but superhuman), mechanized though not (yet) mathematicized? And then, as fashions changed,

> the innovations of Darwin were shaking the foundations of belief. Nothing could seem more opposed than the old doctrine and the new—the doctrine that looked on the world as the work of the divine watchmaker and the doctrine that seemed to draw all things out of chance, chaos, and old time. (Keynes, 1927/1972, p. 276)

where, I note, chaos and old time (though probably not chance) were already associated to (vague and/or Aristotelian) notions of infinity.

Contrariwise, in the mathematical community, the source seems to have been the first fully formalized theories of infinity, as seen next.

Infinity in Mathematics

Dedekind and Peano, with their work on formalizing the natural numbers and other number systems, and their competing definitions of "finite" and "infinite,"[8] both played roles in the formalization of infinity; but while Peano argued (ultimately unsuccessfully) against "completed infinities,"

> Dedekind broke entirely new ground in his free use of "completed" infinite sets as single objects on which one could compute as with numbers, long before Cantor began his work on set theory. [... N]o predecessor or teacher of Dedekind could have inspired this bold step, since they all stuck to the old taboo against the "actual" infinite. (Dieudonné, 1984)

Although the development of axiomatic set theory was neither quick, monotonic, nor painless (Potter, 2004, gives an engaging but challenging treatment, equally mathematical and philosophical; Reid, 1970, tells Hilbert's part of the story biographically), by the 1930s axiomatic set theory had been subsumed into Azzouni's "benign fixation of mathematical practice" (2006, p. 28), and although it had (and has) several variants, it had become entirely standard for mathematicians to make "free use of 'completed' infinite sets as single objects" (Dieudonné, 1984).

Among the axioms of the most common version of set theory today, known as ZFC (Zermelo-Frankel with Choice), two are particularly closely related to the mathematicization (and taming) of infinity.

Informally, the Axiom of Infinity says that the purported characterization of a set **N** proferred above on p. 36 actually does refer to an existing set. In Aristotelian language, it says that there exists at least one "actual infinity by addition" (namely, **N**) in the set-theoretical universe; other axioms then can be used to prove that there are many others, built "by addition" like their prototype **N**. Like the infinitude of time for Aristotle, expressed by saying "there is always a next day," these infinities by addition could also be called "upwards" (or "onwards") infinities.

The Axiom of Infinity can be called either a 'positive axiom'—it asserts that a certain property actually is possessed by some set—or a 'permissive axiom': it provides us with such "fertile conceptual structures and modes of inference" ("Fruchtbaren Begriffsbildungen und Schlußweisen"; Hilbert, 1926, p. 170) that, with it in place, "No one shall expel us from the Paradise [of sets] that Cantor has created" ("Aus dem Paradies, das Cantor uns geschaffen, soll uns niemand vertreiben können"—ibid., my translations).

In contrast, the Axiom of Foundation (or Regularity), introduced by John von Neumann (1925), can be called either a "negative axiom"—it asserts that there do not exist any sets possessing a certain other property (a bit too technical to state here)—or a "prohibitory axiom": It keeps us from falling into the set-theoretical version of the ancient logical sin of "infinite regress" and its more recent versions, the Burali-Forti Antinomy and Russell's Paradox (Mirimanoff, 1917).

The simplest example of a statement about a would-be set X that, when the Axiom of Infinity is adopted, cannot in fact be true for any set X, is "X is a member of X"—symbolically, $X \in X$. Such a set would spawn an "infinite descending $\in$-chain" $\cdots X \in X \in X \in X$, and would-be sets with more complicated descriptions also forbidden by the Axiom of Foundation would spawn more complicated geometries of "endless descent, repetitive or not, with respect to membership" (Quine, 1969, p. 285). By forbidding "endless descent," the axiom denies the existence of what might be called "downwards" (or "backwards") infinities.

In the language of mathematical recursion, the Axiom of Foundation says the hierarchy of sets (in which each set is a level, and level L is subordinate to level M if and only if L is a member of M) is transitive and well-founded.

RECURSION AND INFINITY IN THE HUMAN SCIENCES

In contrast to the situation in mathematics, in the human sciences there simply is no consensus whatever on the meanings and uses of "recursion," "recursive," and so on. In this section I first give a representative sample of these uses, and trace them through a variety of more or less tangled lineages to a few "base cases." Then I repeat this exercise for the meanings and uses in the human sciences of "infinity," "infinite," and so on; there the lineages are fewer and less tangled.

A Recursive Sampler

Each of the following exemplary texts gives an explicit or implicit definition or characterization of "recursion," "recursive," or the like. They are taken from academic publications in all decades from the 1940s to the 2000s, and come from a wide range of human sciences: linguistics ([1], [4], [25]), anthropology ([1], [8], [10], [13], [21]), economics ([2], [14]), psychology ([3], [5], [6], [16], [22]), human geography ([8]), human ecology and consciousness studies ([9], [10], [12], [21]), sociology ([10], [11], [18], [19], [24]), political science ([17]), and education theory ([15], [20], [21], [22]). Italics, bold, etc., follow the originals; translations are mine.

[1] An example of a recursive definition is that for the relation *'ubá·* "father" among the Hausa for West Africa, which we symbolize by '**U**'.

$$(x\mathbf{P}y \cdot x \in \mu) \supset x\mathbf{U}y.$$
$$(z\mathbf{U}x \cdot z\mathbf{U}w \cdot x\mathbf{U}y) \supset w\mathbf{U}y.$$

(Greenberg, 1949, p. 62)

[2] e. The system is *recursive* with respect to *time*, in the following sense: If we know the past of all variables, if the future of the exogenous variables is assumed to be known and if the future disturbances are disregarded, the system gives us recursively the future development of the endogenous variables.

f. In the recursive deduction, the variables are obtained consecutively, one by one, from the explicit relations referring to the endogenous variables. In other words, the system is *recursive* also with respect to the variables. (Wold, 1949, p. 14)

[3] La différence fondamentale entre le raisonnement par récurrence et les inférences logiques bivalentes est, en définitive, celle-ci : la logique bivalente ne connaît que la rapport de la partie au tout et ne détermine la partie qu'en fonction du tout [...]. Le raisonnement par récurrence est, au contraire, un passage de l'élément à la totalité [...]. [The fundamental difference between recursive reasoning and the inferences of two-valued logic is, finally, this: two-valued logic only knows the relationship of the part to the whole and only determines the part as a function of the whole.... Recursive reasoning is, on the contrary, a passage from the element to the totality.] (Piaget, 1949, p. 387)

[4] Moreover, it is possible to show that certain types of concept introductions which look circular are not so in fact—types in which the elimination of the newly introduced term does not involve an infinite regress [...]. [C]oncept formations of these kinds are in regular use in mathematics, and especially in mathematical logic, where they are known as special cases of RECURSIVE DEFINITIONS. (Bar-Hillel, 1954, p. 234)

[5] le propre de la récurrence n'étant pas de répéter le même acte sans plus, mais de répéter un acte qui est à la fois le même et autre chose [the distinctive quality of recursion being, not to repeat the same act and no more, but to repeat an act that is at once the same and different] (Papert, 1960, p. 123)

[6] An attempt would be made to assess the comparative difficulty of two problems which demand the induction of recursive rules. In one the child might be shown, say, a "recursive picture" of a man holding a picture of a man holding a picture of a man holding..., and tested for his ability to grasp the principle of the picture's construction, to recognize that only artistic and visual limitations keep the series from being ex-

tended indefinitely. (Flavell, Botkin, Fry, Wright, & Jarvis, 1968, p. 224; aposiopetic ellipsis points in the original)

[7] We become self-conscious through self-observation; by making descriptions of ourselves (representations), and by interacting with our descriptions we can describe ourselves describing ourselves, in an endless recursive process. (Maturana, 1970, p. 5)

[8] *recursive* behaviour, that is any repeated form of activity in which the results of previous experience—and consequently learning—are taken into account (Walmsley, 1973, p. 50)

[9] [T]here seem to be two species of recursiveness, of somewhat different nature, of which the first goes back to Norbert Wiener and is well-known: the "feedback" that is perhaps the best known feature of the whole cybernetic syndrome. The point is that self-corrective and quasi purposive systems necessarily and always have the characteristic that causal trains within the system are themselves circular [...]. The second type of recursiveness has been proposed by Varela and Maturana. These theoreticians discuss the case in which some property of a *whole* is fed back into the system, producing a somewhat different type of recursiveness [...]. We live in a universe in which causal trains endure, survive through time, only if they are recursive. (Bateson, 1977, p. 220)

[10] C'est donc un processus récursif : tout processus dont les états ou effets finaux produisent les états initiaux ou les causes initiales. [It is therefore a recursive process: any process whose final states or effects produce its initial states or causes.] (Morin, 1977, p. 186)

[11] The recursive character of language—and, by generalization, of social systems also—cannot be understood unless we also understand that the means whereby such systems are reproduced, and thus exist as systems, contain within them the seeds of change. 'Rule-governed creativity' is not merely (as Chomsky's linguistics suggests) the employment of fixed, given rules whereby new sentences are generated; *it is at the same time the medium whereby those rules are reproduced and hence in principle modified.* (Giddens, 1979, p. 18)

[12] *Lineal* describes a relation among a series of causes or arguments such that the sequence does not come back to the starting point. The opposite of *linear* is *nonlinear*. The opposite of *lineal* is *recursive*. (Bateson, 1980, pp. 250–251)

[13] The formal mathematical model developed here is compatible with theories based either on the existence of kinship-defined marriage rules or on the exchange of women between sociological groups. Hence marriage rules expressed in kin-type notation can also be recursively defined in terms of wife-givers or wife-takers of ego's group. (Tjon Sie Fat, 1981, p. 387)

[14] Recursive relationships are unidirectional whereas nonrecursive relationships are bidirectional or cyclical. (Fornell, 1984, p. 9)

[15] The authors discuss 3 versions of cybernetics (C-I, II, and III). They claim CIII can be achieved by recursive, complementary switching between CI and CII modalities. They claim that this switching is a natural occurrence in autopoietic living systems, as defined by M. & V. CI = control via negative feedback, CII = extension via positive feedback. They claim evolution is a recursive interplay between these two. (Sawada & Caley, 1986, as abstracted and summarized by Palmer, 1996, p. 1122)

[16] "recursion," the process whereby the mind or a computer program loops back on the output of a prior computation and treats it as a given that can be the input for the next operation (Bruner, 1986, pp. 96–97)

[17] recursive (i.e., programmed by an external agent) (Dobuzinskis, 1987, p. 52)

[18] It is not only that each position is based on a paradox, but that together these positions interact and produce a further paradoxical situation which, in effect, is a double bind. Furthermore, attempts to resolve this problem in favour of either approach lead to each analysis getting trapped within the recursive loop. (Doran, 1989, p. 523)

[19] recursive, in the sense that subsequent actions may affect previous ones, thereby having either positive (exaggerating) or negative (inhibiting) effects (Richmond, 1993, p. 9)

[20] In the humanities, however, recursion has a broader meaning. Here, it refers to the act of a mind or self "looping back," "turning around," or reflecting on itself, and in this way actually creating itself as a conscious self—the highest expression of human awareness. (Doll, 1993, pp. 288–289)

[21] The definition of a relation satisfied by an entity A: A = {......} is "recursive" if and only if:

a. (*Self-reference*) there is an element of permanence (circularity) in it: its right hand side contains, in some way, the entity A that is present in the left hand side;
b. (*Level-stepping*) there is an element of change (transformation) in it: sterile infinite regress—the danger of all impredicative definitions of type a.—is avoided by the entity being present, in the right hand side, in a somewhat different way from the left hand side. (Sawada & Caley, 1993, p. 3, stating their modification of a definition of Vitale, 1989)

[22] recursive [...] everything may causally affect everything else (Harvey, 2000, p. 31)

[23] recursion is not an act of a mind or a self, but is, rather, an ontological characteristic of the being of the inheritances entrusted to us as teachers and students [...] rather than recursion being an act of the mind, or of a self "looping back" or "turning around," things themselves have a way of turning on us, demanding things of us, laying claim to our attention. (Clifford, Friesen, & Jardine, 2001, p. 3)

[24] It is clear that the three treatments—by myself, Goffman, and Lefebvre—make mutual awareness recursive, since they involve repetitions of awareness of awareness. (Scheff, 2005, p. 375)

[25] Recursion: The embedding at the edge (tail) or in the center (nested) of an action or object [...] of the same type. The output of the current embedded action or object will become the input to the next. (Kinsella, 2010, p. 183)

A Tower of Babel, a Confusing Braid, a Handful of Roots

Even a casual look at examples [1]–[25] show they are not just different ways to say the same thing—in fact, some contradict others: compare [14] (an economist discussing "multivariate methods" for marketing research) with [19] (a sociologist describing a "systems model" of migration) and [22] (a social psychologist describing ways to study "loss and trauma" through narratives). This confused state of affairs has been noted before, even within single human sciences. In the late 1980s, a sociologist of law complained

> In the context of causal analysis, as carried on in empirical research (e.g., path analysis) *non*recursive models are employed, to denote the case of *mutual* influencing of variables. When the autopoiesis literature speaks of recursive processes, it is presumably those nonrecursive models of causal analysis that are meant. What a tower of Babel! (Rottleuthner, 1988, p. 119, italics in the original)

And a physicist working in cognitive psychology and education theory wrote

> One is led to wonder if all authors are talking about and experimenting with the same notion and, if not, what this notion could be. As it happens, a careful reading shows that it is not so and that, unless a very loose and rather useless definition of the term ["recursive"] is assumed, it could be worthwhile to separate this confusing braid into its constituent strands [...]. (Vitale, 1989, p. 253)

More recently an evolutionary linguist observed that "Definitions of recursion found in the linguistics and computer science literatures suffer from inconsistency and opacity" (Kinsella, 2010, p. 179), and a political scientist, having noted that "the term 'recursive' [...] has multiple uses in the political science literature," proposed addressing "[t]he problem of divergent meaning [...] through a survey of potential for reconciliation or possible substitute terminology" (Towne, 2010, p. 259)—a program that may be feasible for political science (a field in which, after all, at least some practitioners presumably have professional insight on "reconciliation"), but not one that I think has any chance of success for the human sciences taken all together.

I have been able to trace almost all the divergent meanings documented in examples [1]–[25] back to specific base cases.

[α] In a few examples, the base case is the pre-modern, informal but still thoroughly mathematical meaning associated with **R1** and **R2** alone.

- The use of "recursive" in economics and allied fields, as by Wold (1949), example [2], falls into this category, although the history is somewhat confused. Flood (1938) uses "recursive" to refer to "methods" that he credits to papers of Tinbergen (1935) and Wold (1938). However, the word itself appears in neither of those sources. Instead, both articles describe what their authors call "complete" systems of equations (or "relations"). "[T]he pertinent thing is that a complete system may be reduced to a single relation involving but one of the fundamental variables" (Wold, 1938, p. 187); that is, a "complete system" is one that allows "reduction to a base case." (Wold credits this idea of "completeness" to Frisch, 1933.)

  As a matter of fact, Flood was trained as a pure mathematician, not as an economist, and actually had used "recursive relation" in just this (not fully formalized) way in his PhD thesis (1935, p. 855), so it is quite possible that he may have been the one to introduce the word "recursive" to economics, as his own description of a concept already introduced by Tinbergen and Wold (and Frisch). On the other hand, I have found no indication either that Wold ever cited Flood, or that Wold ever used the world "recursive" in print before 1949. In particular, Wold nowhere uses the word "recursive" in the first edition of his thesis (Wold, 1938)—although in his "Notes to the second edition" he writes "The scheme (320) falls under the heading of the recursive process" (Wold, 1954, Appendix 1, p. 195; in both editions, "scheme (320)" appears on p. 187 directly above the phrase quoted earlier). Presumably Wold (whose doctorate, like Flood's, was in mathematics) was as familiar as Flood with the informal mathematical use of the word, and his use of "recursive" almost surely arose from the premodern mathematical use.
- Fornell's "multivariate methods" paper, example [14], also belongs to this lineage. Many more examples of this lineage exist outside economics, wherever practitioners find it appropriate to adapt econometric models (and their underlying assumptions about causality) to their own purposes. Alexander and Pallas (1983, pp. 32–33) provide a typical instance from sociology:

> the status attainment process [...] posits a recursive, step-wise flow of influences, beginning with family background characteristics and proceeding, in sequence, to measures of school performance and standing, to interpersonal supports, and, last, to subjective dispositions as determinants of later educational and occupational attainments.

[β] In several more examples, the base case is the modern mathematical meaning formalized by Dedekind along the lines of **R1–R4** above.

- Greenberg (1949), example [1], very openly adopted his use of "recursive" from mathematics (specifically from mathematical logic and the mathematical theory of "relations"), though the word itself does not appear in the only relevant text he cites, Carnap (1929). However, in Greenberg's later work, "recursion" and such words barely appear, and I have found no evidence that any other author's use of such words is due to Greenberg's writings (of course oral transmission or transmission via unpublished writings cannot be ruled out in this and similar cases).
- Bar-Hillel (1954), example [4], was equally forthright about the origins of his use of "recursive": he described it as "the regular use in mathematics." A year later, in his review of Bar-Hillel (1954), Chomsky (1955, p. 45) was agnostic on the question of the potential usefulness of "the insight and techniques of logic"—in particular, of "recursive definitions"—in linguistic theory. Two years further on, Chomsky had become a convert: "If a grammar does not have recursive devices [...] it will be prohibitively complex" (Chomsky, 1957, p. 24).

  Chomsky's use of "recursive" (there and thereafter) is phrased differently from the regular use in mathematics, but is essentially the same. It is reasonable to trace contemporary linguists' uses (e.g., example [25]) back to the modern mathematical meaning via Bar-Hillel followed by Chomsky.
- Although example [11] might suggest that Giddens (and sociologists and anthropologists of his school, e.g., Robben, 1989) comes to "recursion" (etc.) via Chomsky, and thus that his base case is modern mathematical recursion, I think he is more appropriately placed under [γ], below.
- The base case of example [13] is uncompromising modern mathematics, including but not limited to modern mathematical recursion. As the author (under)states,

  > The formal model chosen is algebraical [...]. A more elegant model could have been constructed if the elements of the mathematical theory of categories and functors had been consistently applied to the

problem, *but this would probably have detracted from readability.* (Tjon Sie Fat, 1981, p. 380, emphasis added)

- On the other hand, for Bruner (1986), example [16], the base case seems to be modern mathematical recursion via computer programming. Doll (1993) actually quotes Bruner (1986) on recursion in his sentence following that in example [20]. In turn, Clifford et al. (2001, p. 1), example [22], quote "Bill Doll's infamous article (1993, 289) regarding the status of recursion in the humanities." By then, the mathematical content has almost entirely evaporated (or sublimated into postmodernity), but the lineage—direct from Dedekind—remains clear.

[γ] Most of the remaining examples can be traced back to a confluence of the interdisciplinary "cybernetics/information theory" and "general systems theory" movements that flourished from the mid-1940s through, approximately, the mid-1960s (and still exist in some forms) and therefore, in some sense, to (modern) mathematics. However, it does not seem likely (certainly I can find no evidence) that in these examples the specific terms "recursion" and so on were derived from (any) mathematical usages: How the various authors in this group came actually to use those specific terms remains entirely unclear to me—at some point in the early 1970s, those terms simply appear.

"Cybernetics" is associated with, among others, the mathematicians Norbert Wiener (who coined the word) and John von Neumann, who played especially important parts in creating the theory (and practice) of electronic computation and computer programming, and who were key participants in the 1946–1953 Macy Conferences on Cybernetics (about which, see Heims, 1977, 1991). In their published works, von Neumann and Wiener use the words "recursive" and "recursion" very rarely, exclusively in mathematical contexts, and with their strictly mathematical meanings. The context for von Neumann's uses of the words is, precisely, his foundational work on "automata" and the theory of digital (i.e., discrete) computation (as in von Neumann, 1951). Wiener (1948), by contrast, is associated with the theory of analog computation, in which the role played by recursive functions in digital computation is (more or less) filled by so-called "feedback circuits" (he uses "recursive" only in the sense of "definition by induction").

"Systems theory" is associated with the theoretical biologist Ludwig von Bertalanffy (1967)—who named it (and coined the word "equifinality" in that context)—and Anatol Rapoport, a mathematical psychologist, game theorist, early proponent of Peace Studies and so on (cf. his autobiography, Rapoport, 2000). Neither author seems ever to have used the words "recursive," and so on, with any meaning, in their published work.

The terms do appear in works by other adherents of the two movements.

- Example [7], by Maturana, was published as a research report in 1970, in which form it (and its ideas) seem to have circulated fairly widely, before it was reprinted as the first half of Maturana and Varela (1980), the second half of that book being a reprint of the essay "Autopoiesis: The Organization of the Living" (by both authors), written in 1972. The words "recursion" and "recursive" appear in both essays, never defined, but evidently with a skeletally mathematical meaning that retains only **R3**(a)—transitive hierarchical organization—from among the properties **R1–R4**, while ignoring and arguably denying **R3**(b) and **R3**(c)—well-foundedness and being based. Maturana's coinage "autopoiesis" first appears in the 1972 essay, which is cited (without quoting either that word or any form of "recursion") by Morin (1973).
- Gregory Bateson, examples [9] and [12], is a particularly interesting case. He was in on cybernetics from near the beginning, having participated with Wiener and von Neumann in the 1946–1953 Macy Conferences on Cybernetics (about which, see Heims, 1977, 1991). Likewise, he was aware and approving of Bertalanffy's work, writing, for example, "All this, however, has been better and more rigorously said by Bertalanffy" (Bateson, 1960/1987a, p. 234). He credited both movements explicitly:

  > In the last twenty-five years extraordinary advances have been made in our knowledge of what sort of thing the environment is, what sort of thing an organism is, and, especially, what sort of thing a *mind is.* These advances have come out of cybernetics, systems theory, information theory, and related sciences. (Bateson, 1971/1987, p. 321; italics in the original)

  Further, Bateson's name and work are strongly associated with the word "recursive" (as by Harries-Jones, 1995). Yet, as far as I can determine, the letter to John Brockman from which example [9] is extracted (included as an "Afterword" to a collection of essays about Bateson and his ideas, published three years before Bateson's death) is the first time in Bateson's career that he commits the word "recursive" (or any of its derivatives) to print.[9] A few years later, he (nearly) defines his sense of "recursive" in the glossary entry extracted as example [12].
- Similarly, though Morin's 1977 definition of "recursive process" in example [10] is not explicitly attributed to any earlier work, Morin (1977, passim) cites Bertalanffy, Rapoport, Shannon, von Neumann, and Wiener among others. Twenty years later LeMoigne

and Morin (1999, p. 252) explicitly state that "L'ensemble de ces trois théories—théorie de l'information, cybernétique et théorie des systèmes—nous introduit dans un univers des phénomènes organisés où l'organisation se fait avec et contre le désordre." That is: "Together these three theories—information theory, cybernetics, and systems theory—put us into a universe of organized phenomena where organization creates itself with and against disorder" (my translation).

- Like Morin, Giddens (1979, p. 74), in a later chapter of the book where example [11] occurs, acknowledges the influence on "the social sciences" of "'systems theory', which is not clearly distinguishable form 'information theory' or 'cybernetics', all of which have largely arisen outside the social sciences." Like Morin, he does not explicitly attribute his definition of "recursive," though he comes closer:

> the most relevant sources of connection between biological and social theory [...] rather concern recursive or *self-reproducing* systems. There are two related types of theory involved here. One is the theory of automata [...] as modelled in the Turing machine. But this is not of as much interest to the conception of social reproduction as recent conceptions of cellular self-reproduction (autopoiesis) [...]. (1974, p. 75, italics in the original)

- Caley and Sawada (1986) and Sawada and Caley (1986) coined the phrase "recursive complementarity" as part of a "revised description of complementarity" that "privileges the concept of recursion" (Caley & Sawada, 1986, p. 2). The gloss by Palmer (1996), example [15], makes it clear that, as of 1986, Caley and Sawada were in the autopoietic branch of the cybernetic lineage.
- Dobuzinskis (1987), example [17], elsewhere in that book aligns himself with Morin, Giddens, and cybernetics (broadly understood).
- The use of Bateson's phrase "double bind" in example [18] suggests that Doran (1989) belongs in this lineage, although Doran's endnote 12 (p. 259), "For an explicit discussion of recursion, see D. Hofstadter" (referring to Hofstadter, 1979, Chapter 5) suggests at least an informal connection to lineage [β] as well (Hofstadter is a computer scientist).

[δ] The "base case" of all but one the remaining uses appears to be the mathematician Henri Poincaré's paradoxically unformalized and unmathematicized notion of "raisonnement par récurrence," which he promulgated in his philosophical writings, as follows (my translation):

> C'est qu'il n'est que l'affirmation de la puissance de l'esprit qui se sait capable de concevoir la répétition indéfinie d'un même acte dès que cet acte est une fois possible. [It is no less than the affirmation of the power of the mind that knows it is capable of conceiving the indefinite repetition of the same act when this act is possible once.] (Poincaré, 1902, p. 24)

- This expansive view of "raisonnement par récurrence" goes (on the face of it) far beyond the remit of 'mathematical induction' (i.e., the fifth Peano Postulate), which is the usual meaning of "raisonnement par récurrence" (even for Poincaré, in his less philosophical, more purely mathematical writing). It seems to lead directly to Piaget's use of the term in example [3] (see also 1942, where however I could not find as clear a statement of Piaget's own take on recursion), although Piaget alludes to Poincaré's works more often than he quotes them explicitly. In particular, Grize (1960, p. 89), as far as I can determine, was the first in Piaget's research group to publish even an extract (the one quoted above) from what is possibly Poincaré's most ringing endorsement of "raisonnement par récurrence"an epistemological tool.
- Papert's use of "récurrence" in example [5] clearly derives from Piaget, Papert's supervisor for his second doctoral degree, in psychology. On the other hand, Papert's first doctoral degree was in mathematics, and later in his career—in his work on the computer language LOGO, and more generally his work on "constructivism" in mathematical epistemology and pedagogy (all written in his native English)—he used "recursion" in very much its modern mathematical sense (see, e.g., Abelson, Goodman, & Rudolph, 1974; Papert, 1980).
- Example [21]—the modification by Sawada & Caley (1993) of Vitale's interpretation (1989) of Papert's LOGO-based version of "recursion"— is particularly interesting: It shows the authors grafting a nearly but not entirely de-mathematicized version of the modern mathematical meaning of "recursion" (itself already cross-bred with Poincaré's unmathematicized version as transmitted by Papert via Piaget) onto their somewhat earlier cybernetic and autopoietic version (example [15]).

In fact, "level-stepping" as a defining feature of something styled as "recursive" was introduced independently of, and a year before, Vitale.

> One of the best ways to understand the structure of a recursive explanation is to see how it differs from that of a circular one. In a circular explanation, the explanans is a restatement of the explanandum [...]. *In a recursive explanation, the explanans contains a restatement of the ex-*

> *planandum embedded in a larger structure* [...]. A recursive explanation is more informative than a circular explanation just to the extent that we have independent sources of information about the class of causes referred to in the explanation. (Lipton & Thompson, 1988, pp. 219–220; emphasis added)

The late Peter Lipton was a philosopher with a particular interest in the theory of explanations (Lipton, 1991); according to Thompson (personal communication, August 1, 2013), a psychologist with interests in ethology and evolutionary theory, the 1988 coinage was Lipton's, presumably chosen for its connotation of "reaching back." Citations of Lipton and Thompson (1991) seem to have missed the "level-stepping," "reaching back" aspect of their sense of "recursive"; for example, Hull, Langman, and Glenn (2001, p. 514) conflate "iteration (or repetition or recursion, depending on one's terminological preferences)." Nor does their sense for the term appear to have caught on in the philosophical literature (Lipton did not use it in later work). Evidently Lipton and Thompson (1991) fits nowhere in the lineages I am examining.

- The notion of "recursive rules" in example [6], expanded and abstracted further by Miller, Kessel, and Flavell (1970) in a very well known paper to "recursive [...] in the sense of being able to constitute their own domain of application" (p. 614), seems to me distinctly different from Piaget's notion. I put it in this lineage anyway because Flavell clearly does, when he writes that Piaget's

> concept of *vertical décalage* captures the possibility that there are hidden similarities or recursions in children's functioning across different stages of development. "Development in the Piagetian mode has a cyclic character which buttresses the feeling that it is somehow all of one cloth" [...]. (Flavell, 1996, p. 201, quoting Flavell, 1963, p. 408; italics in the original)

This completes the sorting of my examples into lineages by base cases. Table 2.1 summarizes my findings.

Two of my numbered examples do not fit neatly into any lineage.

There is no doubt that Walmsley, example [8], self-identifies as a systems theorist, given his thesis (Walmsley, 1971), a subsequent research report (Walmsley, 1972), and his description in the source of [8] of "General Systems Theory" as "a reasonably well-developed body of thought that aims to unify science by laying the groundwork for a new paradigm of scientific thought explicitly" (1973, p. 51). On the other hand, Walmsley and Day (1972) include Piaget and Inhelder (1967) in their "bibliography and guide to the literature" of "perception and man-environment interaction,"

TABLE 2.1

	Base Case	Human Science	Examples and Practitioners
[α]	**early mathematics**	economics	[2] (Wold, 1949), [14] (Fornell, 1984)
		sociology, etc.	Alexander & Pallas (1983), etc.
[β]	**modern mathematics**	linguistics	[1] (Greenberg, 1949), [4] (Bar-Hillel, 1954), Chomsky (1957), [25] (Kinsella, 2009)
		anthropology	[1] (Greenberg, 1949), [13] (Tjon Sie Fat, 1981)
		psychology	[16] (Bruner, 1986), [22] (Harvey, 2000)
		sociology	[19] (Richmond, 1993)
		educational theory	Papert (1980), [20] (Doll, 1993), [22] (Clifford, Friesen, & Jardine) , Vitale (1989)
[γ]	**cybernetics, general systems theory**	human ecology	[7] (Maturana, 1977), [9] (Bateson, 1977), [12] (Bateson, 1980), [18] (Doran, 1989), Harries-Jones (1995)
		anthropology	[8] (Walmsley, 1973) (?), [15] (Sawada & Caley, 1986)
		political science	[17] (Dobuzinskis, 1987)
		sociology	[10] (Morin, 1977), [11] (Giddens, 1979), Robben (1989)
		educational theory	Caley & Sawada (1986)
[δ]	*raisonnement par récurrence* (Poincaré)	psychology	[3] (Piaget, 1949), [5] (Papert, 1960), [6] (Flavell et al., 1968), Miller, Kessel, & Flavell (1970)
		human geography	Walmsley & Lewis, 1984 (?)
		anthropology	[21] (Sawada & Caley, 1993)

and Walmsley and Lewis (1984) cite Piaget repeatedly (they also, somewhat mysteriously, state that "Shopping is a form of recursive behavior," p. 81). However, whatever the origin of Walmsley's use(s) of "recursive," [8] does not appear to have been the source of any later use in the human sciences outside human geography.

Scheff (2005) presents a more complex case. In paragraphs preceding example [24], Scheff suggests several ways "recursive" and the like might have been introduced to sociology and psychology.

> In one of my own earlier articles (Scheff, 1967), I proposed a model of consensus that has a recursive quality like the one that runs through Goffman's frame analysis [...]. As it happened, Goffman (1969) pursued a similar idea in some parts of his book on strategic interaction [...]. [A] similar treatment can be found in a book by the Russian mathematician Lefebvre (1977), *The*

> *Structure of Awareness* [. . .]. I wonder whether Lefebvre came up with the idea of reflexive mutual awareness independently of my model. He cites Laing, Phillipson, and Lee (1966), a brief work devoted to a recursive model of mutual awareness that preceded Lefebvre's book (1977). However, he also cites his own earliest work on recursive awareness, an article (1965) that precedes the Laing, Phillipson, and Lee book.
>
> It is possible that Lefebvre's work was based on my (1967) model of recursive awareness, even though the evidence is only circumstantial. As Laing, Phillipson, and Lee (1966) indicate, their book developed from my presentation of the model in Laing's seminar in 1964. Since there were some 20 persons there, Lefebvre could have heard about the seminar from one of those, or indirectly by way of others in contact with a seminar member.

In fact, the words "recursive" and "recursion" *appear nowhere in the cited works* by Laing et al. (1966), Scheff (1967), or Goffman (1969). Lefebvre (1977, but not 1965) does use "recursive" (in both the pre-modern and modern mathematical senses, citing Chomsky on p. 27, and thereby placing his uses—or perhaps his translator Rapoport's—in lineage [β]).

Rather, Laing et al., Scheff, and Goffman consistently use the words "reflexive," "reflection," and "reflexivity." These are glossed by Scheff (2005) in a variety of ways: "recursive awareness," "mutual awareness" (harkening back to Goffman's signature phrase, "mutual consideration"; see p. 37), "not only understanding the other, but also understanding that one is understood, and vice versa," "not only a first-level agreement, but, when necessary, second and higher levels of understanding that there is an agreement," and so on. Like Maturana's version of "recursion" (and his and Varela's "autopoiesis"), the relation of these authors' notion of "reflexivity" to mathematical notions of "recursion" is skeletal, retaining from the latter's defining properties **R1–R4** only a part of **R3**.

Yet I am uncomfortable assigning any of these notions (including Scheff's) to lineage [α] (much less [β]), given the absence throughout the cited works of any references to even such a minimally mathematical notion of recursion.[10] Nor, given a similar absence of references to the cybernetic or systems-theoretic notions, can I justify assigning them to lineage [γ] (though Goffman, 1974, p. 40, credits Bateson, 1955/1987, with the notion of "frames"; and of course Rapoport ties Lefebvre to this lineage as well). The ideas of Lefebvre et al. have certainly been influential in the human sciences (for a recent example, see the intriguing application of Lefebvre's theory of reflexive control, 1977, to "intransitivity cycles" by Poddiakov & Valsiner, 2012), but how and when those ideas (from those sources) came to be referred to by words from the "recursion" group, rather than the "reflection" group, is entirely unclear to me (and merits further investigation).

Enthusiasm for, and Horror of, Infinity in the Human Sciences

I can identify four distinct ways that the various English-language discourse communities in the contemporary human sciences use the words "infinite," "infinitely," "infinity," and the like. (Though I have not investigated non-English academic discourse as I did for "recursion," my impression is that it is not much different on this point.) The first three of these usage patterns are based at least implicitly on technical definitions of one or both of the terms, from mathematics, philosophy, and theology, respectively; the fourth is the nontechnical, vernacular usage of the terms. Any of the four can appear with or without (obvious) affect, which when present can be either positive (ranging from calm acceptance of "the obvious" to enthusiasm or even evangelical zeal) or negative (rarely as extreme as *horror infiniti* of a pre-modern sort, but ranging at least from stern dismissal to cool and detached suspicion).

As a starting point for the following discussion of examples, I quote the relevant parts of the entries "infinite" and "infinity" in the Oxford English Dictionary, covering the technical theological and nontechnical vernacular uses of the terms.

> **infinite** [...] A. *adj.* 1. a. Having no limit or end (real or assignable); boundless, unlimited, endless; immeasurably great in extent, duration, or other respect. Chiefly of God or His attributes; also of space, time, etc., in which it passes into the mathematical use (A. 4b). [...] b. In loose or hyperbolical sense: Indefinitely or exceedingly great; exceeding measurement or calculation; immense, vast. [...]. C. n. 1. That which is infinite, or has no limit; an infinite being, thing, quantity, extent, etc. Now almost always in sing. with *the*; esp. as a designation of the Deity or the absolute Being. [...] (infinite, 1900/2013)
>
> **infinity** [...] 1. The quality or attribute of being infinite or having no limit; boundlessness, illimitableness (esp. as an attribute of Deity). [...] 2. Something that is infinite; infinite extent, amount, duration, etc.; a boundless space or expanse; an endless or unlimited time. [...] 3. a. In hyperbolical use (from 1, 2): Immensity, vastness; an indefinitely great amount or number, an exceeding multitude, 'no end' (*of*). [...]. (infinity, 1900/2013)

I omit the OED's technical "mathematical use (A. 4b)" of "infinite" and the corresponding use of "infinity"; both have been obsolete in mathematics for over 100 years. I also omit A. 1d, the sub-entry of "infinite" for the technical philosophical term "infinite regress"; it is a mere list of usage examples.

(T1) The brief account of the full mathematical formalization of infinity on pp. 39–40 contains several technical uses of "infinite" and "infinity" in mathematics. Note 8 on p. 86 gives explicit statements of the two standard technical definitions of "infinite" used in modern mathematics; these are rarely incorporated explicitly into discourses in human (or other) sciences, but they are always implicit in the first kind of technical use of "infinite" and "infinity" in the human sciences.

(a) The mathematical infinities discussed by linguists (and linguistically inclined anthropologists, e.g., Greenberg) are discrete and without serious exceptions[11] denumerable: purportedly, they are—like **N**, the prototypical discrete denumerably infinite set—generated by some ultimately mathematical recursive process via lineages [α] or [β].

> This article [...] explore[s] the expression of a certain kind of cognitive organization in the culture of the Admiralty Islands. In this organization, opposed male and female categories are themselves further subdivided into opposed male and female categories. Such subdivision is infinitely repeatable [...]. This pattern of organization will be called recursive dualism. (Eyde, 1983, p. 3)

> See, Merge is innate but that is almost automatic. You cannot have a recursive system without Merge, or some variant of it. So if there is a recursive system at all, if there is an infinite number of objects, you have to have Merge. (Chomsky, 2009, p. 32)

(b) In all the human sciences, but particularly in psychology, economics, political science, and sociology, mathematical infinities—primarily continuous—are regularly imported for use as part of (non-discipline specific) mathematical machinery, e.g., mathematical statistics or measurement theory. It is rare for the infinitude (as contrasted with the continuity) of these 'utilitarian infinities' to be made explicit or otherwise commented on, even (perhaps particularly) in the case of the prototypical non-denumerably infinite continuum **R**, the (structured) set of real numbers. One class of exceptions occurs in the context of continuous scaling and measurement.

> Many 'physical phonetic' variables are continuous, so physically there is an infinitely large range of possible sounds. (Sampson, 1974, p. 236)

> Obviously, most human behavior is distributed along an infinite continuum reflecting subtle gradations of individual differences. (Clarke, 1981, p. 89)

At least one of the many logically equivalent mathematical constructions of **R** can, in fact, be correctly described as "recursive" in the pre-modern mathematical sense: every real number x between 0 and 1, inclusive, can be "pinned down" by infinitely repeating the process of bisecting an interval (into "left" and "right" halves, not into "male" and "female" categories as in the culture of the Admiralty Islands)—yielding, in effect, the base-2 representation of x. I have found no examples in the human science literature where this (or any other) connection between continuous infinities and recursion has been drawn.

(T2) Philosophers have used the phrase "infinite regress" technically in two quite different areas: the theory of arguments (or proofs, analyses, etc.) and the theory of causality; or, in fancier language, epistemology and ætiology (I use the latter spelling to help keep in view that this use of the word is broader than the use in medicine). In either context, but particularly the first, it may be deemed either "vicious" (in which circumstance it is often called "the problem of infinite regress") or "virtuous." Thus, for arguments:

> There are three strategies for incorporating the epistemological requirement that justifying beliefs be justified. [...] (Brink, 1989, p. 290)
>
> I conclude that the standard objection to strategy (1) contained in the regress argument fails. An infinite regress of justification need not be vicious. (Brink, 1989, p. 295)

and for causes:

> (2) *The regress is infinite but virtuous.* [...] Suppose that event A has a prior cause B, B a prior cause C, and so, perhaps, *ad infinitum.* Few modern philosophers would consider this latter progression to infinity a vicious one. (Armstrong, 1973, p. 254, italics in original)

Maurin (2007) gives a general survey of the situation in philosophy.

In the human sciences, I do not always find it clear whether any given use of "infinite regress" is meant to be epistemological or ætiological. In any case, the term often has vice or virtue ascribed to it, if only indirectly.

(a) Years before he starts to use the word "recursion," Bateson writes of "infinite regress" in the context of "contexts" in psychology and human ecology, and, if he does not embrace it (it is "shocking"), certainly does not reject it out of hand (it may be what "must [...] be studied").

> Even more shocking is the fact that our hypothesis suggests—but does not stand or fall with the suggestion—that there may be an infinite regress of such relevant contexts. [...] (Bateson, 1960/1987b, p. 250)
>
> The observer must be included within the focus of observation, and what can be studied is always a relationship or an infinite regress of relationships. Never a "thing" [...]. (Bateson, 1960/1987b, p. 251)
>
> Or is our view of the world changed when we admit an infinite regress of contexts, linked to each other in a complex network of metarelations? (Bateson, 1960/1987b, p. 271)

Throughout a series of papers on schizophrenia (Bateson, 1950/1987; 1960/1987a; 1960/1987b; 1969/1987; Bateson, Jackson, Haley, & Weakland, 1956/1987), Bateson refers repeatedly to both epistemology (his life-long interest) and etiology—the latter obviously in the medico-psychological sense, but also (as I read him) in the broader sense of the application, in this case to schizophrenia, of a theory of causation, that is, an ætiology. I believe Bateson (1960/1987b) is using "infinite regress" in both senses.

(b) Similarly, Goffman writes of "infinite regress" in an early instance of his "frame analysis" (mentioned on p. 52).

> Each potential source of discomfort for [the stigmatized individual] when we are with him can become something we sense he is aware of, aware that we are aware of, and even aware of our state of awareness about his awareness; the stage is then set for the infinite regress of mutual consideration that Meadian social psychology tells us how to begin but not how to terminate [...] [I]t is understandable that all will not go smoothly. (Goffman, 1963, p. 18)

I read "but not how to terminate" as an expression more of resignation than despair; "that all will not go smoothly" is more despairing (but the despair may be contextual, particular to just this instance of a reflexively—or, on the account of Scheff, 2005, "recursively"—generated "infinite regress" of "discomforting" frames). Again similarly to Bateson, Goffman here (I am quite sure) uses "infinite regress" both epistemologically (the regress begins with "awareness" of "discomfort," certainly a kind of knowledge) and ætiologically (after the first stage, each new awareness is itself a new "potential source," that is, cause, "of discomfort").

(c) Flavell et al., earlier in their book on "the development of role taking and communication skills in children" (the source of example [6], about a child's "ability to grasp the principle" of "recursive rules" by which a "'recursive picture'" can be constructed) include in a footnote:

> One wonders at what point in his role-taking development the child does in fact first become aware of this "infinite regress" property of such role-playing activity ("I think that he thinks that I think," etc.); its achievement strikes one, intuitively, as an important milestone in the ontogenesis of social cognition. (Flavell et al., 1968, p. 53 n.)

Clearly these authors regard the child's "achievement" of awareness of "this 'infinite regress' property" with favor, and take it to be closely related to, possibly generated by, "recursive rules" from lineage [γ]. As with Bateson's and Goffman's, Flavell et al.'s "infinite regress" seems to me to be both epistemological and ætiological.

(d) Sawada and Caley (1993), example [21], take the view that by ensuring that "level-stepping" is incorporated into "recursive" definitions, "sterile infinite regress—the danger of all impredicative definitions" involving self-reference "is avoided." Clearly, for them, negative affect attaches to "sterility," not to "infinite regress" per se.

(e) A familiar relative of "infinite regress" in epistemology is "circular reasoning."[12] Its analogue in ætiology, "circular causality," is less familiar but far from unknown, particularly in the human sciences, where it is closely associated to the cybernetics and systems theories traditions (see, e.g., Bertalanffy, 1967; von Foerster, 1952). Circularity appears in example [9], from Bateson (1977), where "causal trains within the system," which "are themselves circular," are associated with "two species of recursiveness, of somewhat different nature" (namely, cybernetic and autopoietic). Bateson views this circular causality with unambivalent favor: "We live in a universe in which causal trains endure, survive through time, only if they are recursive" (p. 220).

(f) Unlike Bateson (1977), Doran (1989) appears to view circular causality with unambivalent disfavor. He doesn't claim just (as in example [18]) that "reflexivity and recursion" lead to "paradox" and "a double bind," but much more: "Because these sociologists" (in the "Reflexive Camp" of sociology of science) "choose this framework to work in, they are doomed to travel continually along these recursive loops" (p. 517); and, again, "Whatever they do, they are doomed to stay trapped within a loop of their own making, one which could continue *ad infinitum*" (p. 523). Doran's prescription for averting doom is "breaking frame"—that is, "getting out of the 'constructionist' framework itself" (p. 517).

(g) Like Doran, Scheff (2005, p. 384) finds "infinite regress" of "contexts" a potentially problematic consequence of "Goffman's recursive layers of frames." Without naming names, he imputes distress at that prospect to

> [m]any scholars of the human condition [...] Indeed, among scholars in the humanities, it is often taken for granted that contexts involve infinite regress: "Everything is the context for everything else" (Scheff, 2005, p. 384).

Unlike Doran, Scheff sees a way out of the bind.

> This is an empirical, not a conceptual, problem. Context can be defined in an orderly way, enabling the representation of the least numbers of levels of frames and awareness that are needed to make valid interpretations of strips of discourse. (Scheff, 2005, p. 384)

> To my mathematician's eye, Scheff's very sensible suggestion is a manifestation of the humdrum but (as it turns out, over and over again, in mathematical practice) immensely useful observation that, although the set **N** of natural numbers is mathematically infinite, each individual natural number $n \in \mathbf{N}$ is itself finite—precisely because of the "orderly way" that **N** is constructed by recursion.

(T3) Many theologians (writing of various religions and/or theologies, e.g., the Abrahamic religions, Zen Buddhism, and Hinduism) use "infinity," "(the) infinite," and so on with technical meanings; those given by the OED are most representative of (some strains of) Christian theology. Psychologists, sociologists, and anthropologists, studying individual, group, and cultural phenomena around religion and religions, naturally have frequent occasion to mention those extradisciplinary technical meanings in the course of their own intradisciplinary work; for an obvious example, James (1908) quotes or summarizes nearly two dozen personal accounts that touch on "the infinite" in religious experience, all without ever using (or trying to define) the term in his own right.

As far as I can determine, when the terms "infinity" and so on actually are used in the human sciences with technical meaning(s) proper to theology (or some approximation thereto), it is mostly in psychoanalysis and psychotherapy, and mostly approvingly, as by Stein (2011, p. 402):

> Each finite person is a suitable subject for reception of the infinite [...] Everyone is called to incarnate the infinite in this life; everyone is capable of receiving intuitions of the Divine and of finding ultimate meaning in life as lived concretely in each and every human context on the planet.

(~T) Nontechnical, vernacular uses of "infinite" and so on appear throughout academic discourse, including in all the sciences, philosophy, and the humanities, and even in paramathematical discourse (excluding fully mathematical discourse produced by mathematicians for mathematicians, where they have become taboo).[13]

Mutatis mutandis, the 1900-vintage descriptive definitions from the OED appear to fit these uses, as in the following representative examples from a monograph, a textbook, and two journals.

(a) [We] might examine revolutionary movements, fraternal groups, religious revivals, agrarian revolts, and an infinite variety of other movements. (Toch, 1965, p. 86)

(b) In addition, the notion that America was a frontier society and indeed an empty land also created a sense of endless space and infinite opportunity [...]. (Turner, 1999, p. 226)

(c) Although in our counterfactual musings there are an infinite number of ways we can mutate antecedents so that an outcome would be different, we tend to introduce relatively minor mutations that are systematically constrained [...]. (Spellman & Mandel, 1999, p. 121)

(d) Out of the infinity of relationships that could possibly obtain between an infinity of potential variables, the sociological theorist has little choice but to rely on [...] understandings of "what people are like"—in other words, psychology—to choose among them. (Marshall, 2008, p. 19)

A large group of vernacular uses is in phrases like "nearly [almost, practically] infinite," "a near [almost an] infinity," etc. Mathematicians like to say that such phrases are meaningless; and so they are, in terms of the semantics of the consensus technical definition(s) of "infinity" that have been standard in mathematics since the formalization of mathematical infinity described earlier. Kasner and Newman (1940, p. 34) give a canonical mathematician's statement of why never to use phrases like "nearly infinite":

> Above everything, we must realize that "very big" and "infinite" are entirely different. [...] There is no point where the very big starts to merge into the infinite. You may write a number as big as you please; it will be no nearer the infinite than the number 1 or the number 7.

But such pedantry misses a point about the pragmatics of such phrases that is obvious but rarely stated: "'*nearly* infinite' just means that it has *nearly* the same consequence from the relevant perspective as if the thing were infinite" (Dickinson, 2006; emphasis added). This is true even for a writer or speaker who, fully understanding mathematicians' meaning for the word infinity, still does not feel constrained by that understanding (perhaps because she or he is not at that moment playing the social role of mathematician). In short, although Kasner and Newman's point is correct in terms of the meaning of "infinite" to mathematicians, it is neither correct nor incorrect, but simply irrelevant, in terms of the useful functions that phrases like "nearly infinite" (etc.) play in vernacular discourse.

FINITIES AND INFINITIES IN HUMAN EXPERIENCE

In this section I consider 'infinities', and the 'finities'[14] with which they contrast, in a context of human experience broader than the human sciences alone; I take an axiomatic approach, based in "evolutionary ontology."

Approaching (In)finity in the Light of Evolutionary Ontology

In his 1977 William James Lectures at Harvard, Donald T. Campbell reminded his audience—comprising philosophers, social scientists, and others (Campbell, 1988b, pp. 23–24)—that "'Ontology' has to do with the referents of knowledge, if indeed there are any," then explained that

> In my usage this will mainly be about the material, energetic world, epitomized by what physics is trying to talk about, but also including the realities, if any, sociology is trying to describe. [...] (Within modern uses of "ontology" there is one that I will neglect without rejecting. That is the use of ontology to refer to the objects of mathematical and logical discovery.) (Campbell, 1988a, p. 440)

Like Campbell, I am (more than merely) open to the usefulness—I would say, the necessity—of simultaneously considering multiple, not obviously compossible, ontologies when I am trying to talk or think about our given world, in which we all constantly contend (as vernacular speakers and lay speakers, and also as intellectuals and academics who read, write, and proclaim scholarly books, articles, and lectures) with vast arrays of seeming facts of "physics," "sociology," and much else.

Unlike Campbell, I believe it is unprofitable, even impossible, to exclude "the objects of mathematics and logical discovery" in their entirety from these ontologies. Quite the contrary: following Lorenz (1941/1962), I am convinced of the truth of the following axioms:[15]

EO1 Mathematics and mathematical modeling, no less than other "epistemic activities, such as learning, thought, and science" (Campbell, 1974, p. 413), have evolved alongside human biology and sociality.

EO2 The (evolved) ontology of mathematics—and of mathematical modeling, the human activity in which specifically mathematical "referents of knowledge" serve as tools and methods to the general end of being more "effective in thinking about the external world in which we have evolved" (Kreisel, 1978, p. 86)—is inseparable from the evolved ontologies of those other "epistemic activities."

Rudolph (n.d.; see also Rudolph & Valsiner, 2012, especially pp. 24–27) gives the beginnings of a detailed discussion of this "evolutionary ontology" (briefly, EO) that I propose as an obviously useful and potentially valuable counterpart of the "evolutionary epistemology" (EE) developed by Lorenz, Campbell, and many others.[16]

Although there may be reason to believe that the admission into physical scientists' ontologies of 'uncountably infinite' mathematical structures[17] like **R** (the so-called "real number line") is a price worth paying for what they (on this account) contribute to "the unreasonable effectiveness of mathematics in the physical sciences" (Wigner, 1960), it is entirely unclear (to me) whether that ontological commitment (or even the much weaker commitment to denumerable infinities) is worth the while of the human sciences—particularly since there is no consensus that mathematics has (yet) been even reasonably effective in those sciences taken all together.[18]

Note that my skepticism sketched in the preceding paragraph need not (and, I think, does not) conflict with my beliefs about EO sketched in the paragraph before that: Distinct epistemic activities (across species, within the human or other species, within human cultures, within variously constituted human groups, or even for a single human being in different contexts) may call for distinct ontologies (or different pluralities of ontologies); in particular, the human sciences (or some larger subgroup of the "life sciences") may call for different ontologies than "the physical sciences."

This last idea is not novel. Wigner suggested it in the same lecture that first touted his slogan "the unreasonable effectiveness of mathematics in the physical sciences":

> A much more difficult and confusing situation would arise if we could, some day, establish a theory of the phenomena of consciousness, or of biology, which would be as coherent and convincing as our present theories of the inanimate world. [...] [I]t is quite possible that an abstract argument can be found which shows that there is a conflict between such a theory and the accepted principles of physics. [...] Such a situation would put a heavy strain on our faith in our theories and on our belief in the reality of the concepts which we form. (Wigner, 1960, pp. 13–14)

Indeed, well before Wigner, von Uexküll (1920) had given a beautiful affirmative statement of principles intended precisely to "establish a theory of the phenomena" of (what he called) theoretical biology, "which would be as coherent and convincing as our present theories of the animate world."

> The world of the physicist consists (1) of places, the number of which is infinite, (2) of movements, the extent of which is unlimited, and (3) of moments, having a series without beginning or end. [...] The biologist, on the other hand, maintains that there are as many worlds as there are subjects. [...] The subjective world consists (1) of places, *the number of which is finite,* (2) of movements, *the extent of which is limited,* (3) of moments, *in a series that has both a beginning and an end,* and (4) of content-qualities, *which are also fixed in number,* and have laws which are likewise laws of Nature. (von Uexküll, 1926, p. 70; emphasis added)[19]

On this point, I differ from Wigner—I doubt that "such a situation" would put an insurmountably "heavy strain" on *my* epistemological and ontological theories or beliefs—and align myself, at least for purposes of exploration, with von Uexküll. As Rudolph and Valsiner (2012, p. 27) have put it,

> If, in fact, mathematics has so far been relatively ineffective in the social sciences, perhaps it is because 'we' (not just *Homo sapiens sapiens* or even—say—*Hominidæ* or *Mammalia*, but all life) have evolved in the "external world" of physics, chemistry, and life-in-general for so much longer than in the "external world" of human sociality, that is, human life-in-particular. Yet we (humans) need not simply give up on the attempt to make it effective.

With that in mind, I adopt von Uexküll's sentence that begins with "The subjective world" as the Axiom of Subjective Finiteness; it is clearly more foundational than **EO1** or **EO2**, and accordingly I label it **EO0**.

In the next two subsections I launch preliminary investigations, informed by axioms **EO0–EO2**, of finities and infinities in human experience.

"As If" Behaviors and Subjective Finiteness in Human Experience

As I use the term, a subject's "experience" (in particular, the "experience" of a human being or a group of human beings, possibly extended in space, in time, or in both: I use *human subject* to cover all these cases) is no more and no less than that subject's experience of *its subjective world* in the sense of von Uexküll (1926), suitably generalized,[20] and is itself 'in'—better, 'of'—that world (no regress is required). Operationally, a subject's subjective world is that subject's ontology, and this, with von Uexküll's axiom **EO0**, implies that each subject's ontology is finite. Further, a subject's subjective world can (only) be deduced from that subject's behavior, broadly defined; thus the experience and ontology of a subject can, in turn, (only) be deduced—whether by another or by that subject (cf. Laird, 2007)—from that subject's behavior. In particular, for me (as, in different ways, for Meinong, 1904; Mally, 1911; and other object theorists), human acts of all kinds, including "mental acts" like remembering and imagining—if, indeed, the former can be distinguished from a special case of the latter (cf. Bartlett, 1932, p. 214)—necessarily have ontological implications.

I share with Sarbin (2004, p. 9), and for the same reason, a strong "preference for the gerund 'imagining'" (as well as, in my case at least, other verbal forms of "imagine") "rather than the substantive 'imagination.' The gerund 'imagining' connotes an active process, something the imaginer *does*. In contrast, the substantive 'imagination' suggests a thing-like entity, or a property of the mental apparatus [. . .]."

I also find Sarbin's "semiotic analysis [. . .] of the word 'imagining'" (p. 10) very compelling. It leads him to an account of "Imagining as 'As If' Behavior" derived from a more general "thought model available to students of silent and invisible processes" proposed as an "alternative view" to "[t]he contemporary form of Cartesian mentalism." In this "thought model,"

> human beings construct their worlds, within limits. [...] Human beings' constructions of reality depend upon a skill to function at various levels of hypotheticalness. [...] [A] hierarchy of hypothesis-making skills liberates human beings from the constraints of the immediate environment. With this "as if" skill, the actor can interact with narratively constructed events that are spatially distant and temporally remote, he or she can relocate self to different times and places. (p. 11)

This model's "hierarchy of hypothesis-making skills" arises out of a "three-stage sequence of child development" beginning with (i) "outright copying of the performances of another person" who is present, proceeding through (ii) imitation of an absent other, and—"concurrent with [...] the muting of speech"—culminating in (iii) "muted and attenuated role-taking—[...] the ultimate referent for the word 'imagining'" (p. 12).

On Sarbin's (very much standard) account of development, this hierarchy "makes possible the formation of narrative plot structures," his interest in that chapter. My interest here is different, so I would add the following to his hierarchy (without making a general claim about development): (iv) for at least some human subjects, the "'as if' skill" extends to the "formation" of "structures" that are *neither spatially nor temporally located* (and thus are perhaps not reasonably to be called "events"), and from which not only the "actor" *but every other animate subject* is apparently *personally absent* (and thus are perhaps not reasonably to be called "narratives": though if such a structure serves an ætiological function it can, I think, reasonably be said to involve a "plot"). In particular, I assert the following proposition.

> **P:** *The formation and use of many (if not all) mathematical "structures" are best understood as paramathematical "as if" behaviors of type (iv).*

In the rest of this subsection, I argue for Proposition **P** in the special case of certain mathematical finities.

Paramathematical Ellipsis and "As If" Behavior

In the previous section I discussed a pair of distinct rhetorical functions performed by ellipsis points "..." in paramathematical writing and by certain verbal equivalents in paramathematical speech.[21] To this pair of rhetorical functions corresponds an analogous pair of "as if" functions. I describe the first, concerning finities, here, and the second, concerning infinities, in the following subsection.

In the earlier discussion, I showed that the rhetorical function of medial ellipsis points in paramathematical written expressions like "1, 2, . . . , 10" or "*m*, . . . , *n*," and of a few set phrases in paramathematical spoken expressions like "*from* 1 *to* 10" or "*from m to n*," is a form of ellipsis in the classical sense recorded in the OED (ellipsis, 1891/2013). The corresponding "as if" function is exercised by the writer/speaker substituting for an *explicit manifestation* of some mathematical structure (e.g., the collection of natural numbers between 1 and 10, or *m* and *n*) an *implicit reference* to some terminating mathematical process (imagined by the writer/speaker as undertaken by the reader/listener—who may be that writer/speaker) that *would produce* that structure in its entirety, then terminate: that is, the writer/speaker and the cooperating reader/hearer behave "as if" the structure exists, by (implicitly) agreeing to imagine how it *might be produced* in the world(s) of human experience (possibly though not necessarily modified by imagining also that certain limited resources have been greatly—but only finitely—extended).

My use of the modal verbs "would produce" and "might be produced" further points up the "as if" nature of the implicit process (and so, all the more, the "as if" nature of the process-produced structure)—a process that, in fact, is not undertaken by the writer/speaker at the time of writing/speaking, may never have been so undertaken, may never be so undertaken, and may even be physically impossible for anyone ever to undertake. For example, very many people (including children in school or at play) on occasion actually count from 1 to 10; for them, to imagine a terminating process that would produce the set of natural numbers between 1 and 10 it suffices (I do not claim it is necessary) to remember—that is, simultaneously to imagine *and* to assign "veridicality" and "pastness" to that imagining—such a verbal (spoken or written) counting-out or some gestural equivalent. Fewer, but still very many, people have counted from 1 to 1000. No person has counted from 1 to $10^{10^{100}}$ in any world of human experience, and no one ever can: $10^{10^{100}}$ is hugely greater than the number of breaths that will ever have been breathed by all humans who have ever lived. But many can imagine some person doing so (or so *I* imagine).[22]

Formation and Use of Mathematical Finities: Small Natural Numbers

In my statement of Proposition **P**, I took care not to claim what I do not think is true: that the "formation and use" of *all* mathematical structures either are, or are "best understood" to be, paramathematical "as if" behaviors. That care motivated my disclaimer in the previous subsection of any *necessity* for "as if" behaviors (like imagining or remembering) to play a role in formation and use (by a typical appropriately acculturated human subject) of "the set of natural numbers between 1 and 10" and, not incidentally, of each of those ten numbers—or even in a (less typical because more specially acculturated) human subject's formation and use of the analogous verbal constructs in which "10" is replaced by "1000."

There is, in fact, a lot of empirical research into the "formation" of small natural numbers—as signs or "content qualities" (to use the language of von Uexküll) in the subjective worlds of many kinds of beings, not only humans. A representative sample of such research (through 2000), by developmental psychologists, ethologists, and others, has been collected by Lakoff and Núñez (2000, passim but especially pp. 15–22). It demonstrates convincingly (to me) that these instances of formation are ontogenetically prior to anything I would find it reasonable to call "'as if' behaviors."[23]

On the other hand, I know of no empirical research (and find it difficult to imagine any) that supports the proposition that a human subject's formation of, for instance, "the set of natural numbers between 1 and $10^{10^{10}}$" can be understood as anything other than an "as if" behavior. In the following subsections I argue (non-empirically) in support of a contrary proposition: the special case of Proposition **P** stating that the formation and use by human subjects of natural numbers *in general* (that is, with exception of a few cases like those of small natural numbers)—whether individually or in finite collections—truly are best understood as paramathematical "as if" behaviors of type (iv), specifically, as acts of *imagining* certain terminating processes and their products and by-products.

Formation and Use of Mathematical Finities: Large or Indefinite Natural Numbers

The most basic examples of mathematical finities are sequences of consecutive natural numbers, e.g., "1, 2, . . . , $10^{10^{100}}$" or "*m*, . . . , *n*." I claim that a human subject forming and using the first of these example sequences—that begins and ends with the *definite numerals* (i.e., conventional symbolic notations for specific natural numbers) 1 and $10^{10^{100}}$—is acting "as if" s/he has performed or witnessed (or perhaps otherwise become acquainted with) some terminating process more or less like this: the process

- starts with the (already "formed") natural number 1
- proceeds by repeated applications of "the law of ordinal arithmetic" (in the words quoted on p. 36 from Whitrow, 1933), "that every [natural] number has a successor," thus forming (successively) the natural numbers 2 (successor to 1), 3 (successor to 2), *and so on*
- terminates upon formation of the natural number $10^{10^{100}}$

An apparent difficulty with this account—that the imagined process is (in fact) physically impossible of performance or witness by any human agent—is at least somewhat overcome if a further claim is true, that this "as if" behavior is of type (iv), so that the imagined process is neither spatially nor temporally located, nor personally inhabited. (I have no empirical evidence for—or against—either claim, other than my introspections, which favor both; but it does not seem impossible to me that empirical research on the subject could be performed, and perhaps some has been or will be.)

My account of the formation and use of the second example sequence—that begins and ends with *indefinite numerals* (i.e., conventional symbolic notations for indefinite natural numbers)—is a bit more complicated, being a sort of "compound as if" behavior. A human subject forming and using this sequence must first act "as if" *m* and *n* are definite numerals that are already (somehow) "formed," then from that beginning imagine a terminating process along the lines just sketched.

In those and similar examples of mathematical finities, the product of the process is a specified (finite) set of natural numbers; each individual natural number in the specified set is a by-product of the process. But the process is imagined! So, therefore, are its product and its by-products—even in cases (almost surely a vast majority of those that arise in most people's daily lives, though rarer for persons who are much engaged in paramathematical discourse) when these (by-)products also are, or could be, either innate or (truly and justifiably) remembered.

Idealizing as Collective Imagining

My description just above of an "imagined terminating processes" looks very much an instance of the general semi-formal description of a "computation" quoted on p. 29 from Soare (1996, p. 286). This similarity is of course not accidental—but I did *not* write my description as I did because I think it a good or profitable idea ever to call a human "process," least of all an imagined one, a "computation" in the sense of the formal mathematical consensus definition that Soare is describing. Rather, if my account is close to right, the resemblance in our descriptions arises naturally and appropriately because

(1) the *imagined* process that I *describe*, and
(2) the *idealized* activity of a human "computer" described and formalized into an (imagined) "automatic machine" by Turing (1936), later named a (formal) "Turing machine" (Church, 1937), and eventually subsumed into a (formal, terminating) "computation" as summarized by Soare,

have a common basis—namely, *actually experienced* human behaviors such as *counting off* some (short and definite) sequences of consecutive integers. The principal distinction between (1) and (2) (aside from the broader scope of the latter) comes down, I think, to the distinction between two different "as if" behaviors of type (iv): "imagining" and "idealizing."

That distinction, in turn, is chiefly due—as I see it—to the behaviors' belonging to different "levels of organization": "imagining" is a behavior of individual human subjects; "idealizing" can be (and perhaps must always originate as) a behavior of individuals, but tends eventually to become (and persists as) a behavior of an "over-individual" human subject (Überindividuum) in the sense of Münsterberg (1900) and Werner (1959).

> According to Werner, humans are members of an "over-individual" unit through which they receive characteristics that can be understood only through this higher unit. For example, expressions of the "over-individual" unit are seen in language, religion, law, and customs that can never be rooted within the individual alone. What these hierarchically higher placed units have in common is that they are the products of and originate from *collective mental life.* (Diriwächter, 2012, p. 205; italics in the original)

Here, the notion of "collective mental life" is not (and ought not to be taken to be) at all mysterious, but if the reader objects, for present purposes the phrase could be replaced by "culture"—specifically, "mathematical culture" (with its characteristic "benign fixation" identified by Azzouni, 2006).

"As If" Behaviors and Purported Infinities in Human Experience

In this subsection, I argue for Proposition P in another special case, that of certain mathematical infinities.

Paramathematical Aposiopesis and "As If" Behaviors

In the previous section, I argued that the rhetorical function of terminal (or, rarely, initial) ellipsis points in paramathematical written expressions like "1, 2, 3, . . . ," and of a small store of set phrases in paramathematical spoken expressions like "1, 2, 3 *and so on*," has two aspects that correspond to two aspects of aposiopesis in the classical sense recorded in the OED (aposiopesis, 1885/2013): a "rhetorical artifice, in which the speaker comes to a sudden halt, as if unable" (the first aspect) "or unwilling" (the second) "to proceed." To each of these rhetorical functions corresponds, in its turn, an "as if" function.

(1) The "as if" function corresponding to the first aspect of aposiopesis is exercised by the writer/speaker "coming to a sudden halt" *because* (not "as if"!) "unable to proceed." The inability to proceed is not, however, an inability to take the next step, as might cause a walker's "sudden halt" at a physical barrier: It is an inability to take *all* the next steps, which *might* be recognized—suddenly—at any step (unobstructed though it be), and *must* be recognized—eventually—if only by the mortal walker's survivors. By *the act of halting* (signaled

by ellipsis points or a set phrase), in and of itself, the writer/speaker substitutes an *implicit reference* to some *non-terminating* mathematical process (imagined by the writer/speaker as undertaken by the reader/listener—who may be that writer/speaker) that *would produce* some specific mathematical structure, were the process—counterfactually—to *proceed forever,* and were the reader/listener to be present "after forever" to witness that product. That is, the writer/speaker and the cooperating reader/hearer behave "as if" the structure exists, though what they *do* is (implicitly) agree to imagine how the structure *might be produced* in a world of (necessarily counterfactual) human experience.

(2) The "as if" function corresponding to the second aspect of aposiopesis is like that in (1), except that even an implicit reference to any "non-terminating computation" (mathematical process) may be absent, as may be the specificity of whatever (mathematical) structure would be its product.

Formation and Use of Mathematical Infinities

In contrast to the situation described in the second paragraph under the heading "Formation and use of mathematical finities: small natural numbers" on p. 65, there is—to the best of my knowledge—no empirical research that suggests any innate "formation" of mathematical infinities by human (or other animate) beings, as signs, "content qualities," or anything else, in their subjective worlds or elsewhere.

Even if it could somehow be shown that the Basic Metaphor of Infinity of Lakoff and Núñez (2000, p. 155ff.) is "innate" (to human beings) in some manner (as small natural numbers apparently are), it would not follow that some mathematical infinity (in the sense of note 8) of existents—be they mathematical structures, physical entities, or what have you—can belong to some subject's ontology: at most what would follow would be (what I am sure is true empirically, with no need to derive it from the Basic Metaphor!) the formation in some human subjects' ontologies of certain "'completed' infinite sets *as single objects*" for "free use" (referring back to the matter quoted on p. 39 from Dieudonné, 1984; emphasis added), such as **N**.

That is, I accept that for, at least, certain human subjects (e.g., me and my mathematical colleagues; or, indeed, the present-day "mathematical culture" within which we subsist), **N** and similar mathematical structures comprising a mathematical infinity of substructures do "exist" in an "as if" way, i.e., *"as if" they have been constructed by human subjects* within those subjects' "constructions of reality" (extending the phrasing of Sarbin, 2004, p. 11, beyond individual "human beings"); but I deny that this either implies or is implied by the claim that each of those substructures (e.g., in the case of **N**, each individual natural number) "exists" for those human subjects, *even in an "as if" way* (or, conceivably, in some way that is 'metaphorical' but is not "as if"), and in fact I reject that claim (for every subject).[24]

Rather, my account of the formation and ("free") use of "infinite sets as single objects" is similar to my previous accounts of the formation and use of mathematical finities. The case of **N** itself is representative. I claim that **N** is formed in the act of imagining a non-terminating process—for **N**, the fundamental such process, namely, Whitrow's "law of ordinal arithmetic" or its formalization as Peano's postulates (I)–(IV): that is, the process that

- starts with the (already "formed") natural number 1
- proceeds by repeated application of "the law of ordinal arithmetic," thus forming (successively) the natural numbers 2 (successor to 1), 3 (successor to 2), and so on
- but (in contrast to the earlier case of processes generating large or indefinite natural numbers) does not terminate

The last of these points is (I think) obviously necessary, but is it possible (obviously so or otherwise)? At least one argument has been published—albeit in a different context, and nearly 60 years ago—that may appear to show that specifically this non-terminating process cannot be imagined (and nor, implicitly, can any other).

"But: What Stands 'Etc.' For?"

"Strict finitism" (also known as "ultra-finitism" or "ultra-intuitionism," Van Bendegem, 2010; cf. Ésénine-Volpine, 1961) is a philosophy of mathematics that, *inter alia*, takes an ontological stance questioning or denying the existence of arbitrarily large integers. The paper "Is $10^{10^{10}}$ a finite number?" (van Dantzig, 1955/1989) has been called "one of the best candidates" to be "an original source for strict finitism" (Van Bendegem, 2012, p. 142). In it, van Dantzig, an early student of the intuitionist topologist L. E. J. Brouwer, undertakes to show that Brouwer's "intuitionistic mathematics can not be said to be absolutely 'exact'" (throughout, italics and paragraph numbers are van Dantzig's).

> 8. Brouwer's "Over de grondslagen der wiskunde" (1907) begins with the words (in translation): "One, two, three . . . ; we know this sequence of sounds (spoken ordinal numbers) by heart as a sequence without end, i.e., continuing itself always according to a known law." If one tries to find out what the dots stand for, one sees that Brouwer's statement can not be maintained . . . ; we do *not* know the "whole" sequence by heart, and it does *not* continue according to a known law. Going on, one arrives at million, . . . , billion, . . . , trillion, . . . , quadrillion, quintillion, sextillion, . . . and—knowledge of Latin getting scanty—, millinnillion, . . . , millionnillionnillion, . . . , millionnilli . . . illion

> (million times repeated), etc..... But: what stands "etc." for? (van Dantzig, 1955/1989, pp. 260–261)

Earlier in the paper van Dantzig appears to dismiss the possibility of a "known law" being imagined:

> 2.[...] Weakening the requirement of actual constructibility [e.g., printing, counting out aloud, or engaging in Brouwerian "elementary mental acts"] by demanding only that one can imagine that the construction could actually be performed—or, perhaps one should say rather, that one can imagine that one could imagine it—means imagining that one would live in a different world, with different physical constants. (p. 259)

Does this dismissal of Brouwer's "known law" as being unimaginable without also "imagining that one would live in a different world" apply as well to my "act of imagining a *non-terminating process*" consisting of "repeated application of 'the law of ordinal arithmetic'"?

I do not think so—and for a reason somewhat parallel to one that van Dantzig enlists against Brouwer just following the matter last quoted.

> 3. The result of 2. seems to be contradictory: it is impossible to construct natural numbers as large as $10^{10^{10}}$, but $10^{10^{10}}$ is a natural number. The contradiction, however, is apparent only, as one has meanwhile unconsciously changed the meaning of the term "natural number." (van Dantzig, 1955/1989, p. 259)

Here van Dantzig dissolves an apparent contradiction by recognizing in it an (unconscious) change of meaning, that is, the fallacy of equivocation. My case is similar, in that (I claim) van Dantzig and I are using importantly different meanings of "imagining," his much more stringent than mine:

> 1. Unless one is willing to admit fictitious "superior minds" like Laplace's "intelligence," Maxwell's "demon" or Brouwer's "creating subject," it is necessary, in the foundations of mathematics like in other sciences, to take account of the limited possibilities of the human mind and of mechanical devices replacing it. (van Dantzig, 1955/1989; for some relevant discussion of "Laplace's 'intelligence'" and "Maxwell's 'demon'," see Rudolph, 2012b, pp. 294–301)

That is, if I understand correctly, van Dantzig bases his dismissal of the possibility "that one can imagine that one could imagine" an implementation of the "known law" on the incompatibility of two conditions both of which he accepts as necessary:

(vD1) The implementation must be imagined to take place in—and in detailed accord with all the physical laws of—the actual physical universe co-inhabited by the imaginer, his readers, and van Dantzig.

(vD2) The imagining of the implementation must be performed using only "the limited possibilities of the human mind" as it actually exists.

Certainly I accept (vD2), since (for me) both "imagining" and "the human mind" exist as, and only as, (human) behaviors. But I reject (vD1) without reservation; in fact, I reject even the much weaker condition that, merely, some physical universe (van Dantzig's "different world, with different physical constants") be imagined as the locus of the implementation of the "known law." On the contrary, what I have described as "as if" behaviors of type (iv)—including in particular the "as if" behavior of imagining a non-terminating process and thereby imagining that its product and all its by-products are constructed—are distinguished from Sarbin's "as if" behaviors of types (i)–(iii) especially by their disconnection from any (real or imagined) physical universe.

Purported Mathematical Infinities in the Human Sciences

I have just argued that human subjects who use **N** or other mathematically infinite sets do not thereby make ontological commitments to each of a mathematical infinity of 'structures', so the fact of some human subjects' using **N** (etc.) while 'doing mathematics' does not contradict the Axiom of Subjective Finiteness. I contend that, similarly, human subjects who use the various notions of "infinity" (etc.) represented in the example texts under the headings (T1) through (~T) on pp. 54–60 do not thereby make ontological commitments to each of a mathematical infinity of 'structures' (etc.). If this contention is so, then the fact of some human subjects' using such notions of "infinity" (etc.) while "doing human science" does not contradict that axiom. I devote the bulk of this section—all but the next two paragraphs—to arguing for my contention in the hardest case, that of mathematically technical uses like those in the example texts under heading (T1).

No detailed argument seems necessary for technical theological and non-technical vernacular uses, types (T3) and (~T), from which (at least on the evidence of my examples) even an imagined terminating process of counting (or measurement) is absent and no ontological commitments at all, much less infinitely many, appear to be made.

The case of philosophically technical uses, type (T2), is potentially a bit more complicated, in that all contemporary philosophers are surely aware of the technical mathematical definition of infinity. I am not sure how many of them, while doing business as philosophers, allow that definition to impinge on their use of the philosophically technical term "infinite regress," or on its various semantic relatives like "circular reasoning" and "circular causality." However, as I read examples of uses in the human sciences—like those under (T2) and many more—of such terms, I see no evidence that *those* uses either entail, or are intended or understood by their authors (who uniformly give no indication that I can discern of having in mind anything at all like the technical mathematical definition) to entail, ontological commitments to a mathematical infinity of existents of any sort.

I base my arguments for type (T1) on two general observations—partly empirical, partly mathematical, and both running counter to some common intuitions—about mathematically finite sets (with members that need not, however, themselves be mathematical "things"):

(a) many such finite sets in human experience are *humanly uncountable*
(b) any such finite set has (in no fewer different ways than it has members) the topological structure of a *mathematical continuum*, and many such sets in human experience have some such structure that is 'natural' in both mathematical and psychological senses.

I expand on (a) and (b) in turn, illustrating each by applying it to a specific infinity of the sort purported in the example texts under headings (T1a) and (T1b), respectively.

Humanly Uncountable Finities

The only technical mathematical meaning of the term "uncountable," given by Cantor (1874) in the language of set theory, is such that an uncountable set is *a fortiori* an infinite set (see note 8). No term of the form "⟨ADVERB⟩ uncountable" ever been given a technical mathematical meaning, much less one that could apply to a finite set.

Recently, however, E. and H. Luuk have proposed that "physically uncountable" be used as a technical term in linguistics and cognitive science, explicitly stating that "the word 'uncountable' is not used here in its common set-theoretic sense (to mark the difference between countably and uncountably infinite sets)" (Luuk, 2013, p. 90). "[W]e claim that a property of natural language is physically uncountable finity and not discrete infinity" (Luuk & Luuk, 2011, p. 1).

> [L]anguage is a physically uncountable finite set—it has a finite number of elements (NL expressions) that cannot be physically counted (Luuk & Luuk, 2011). Other physically uncountable finite sets include, e.g., the number of atoms in Jupiter or the Solar System at a given moment etc. (Luuk, 2013, p. 90)

That is, they define "physically uncountable" to mean "cannot be physically counted." Their intended meaning for the modified verb "physically count" is not entirely clear to me; I propose to interpret it along the following lines.

(PC1) The core verb *count* is that particular case of *compute* (i.e., to perform a "computation" in the sense of the consensus definition quoted from Soare, 1996, on p. 29) referring to a computation that constructs a one-to-one correspondence between a (computable) set and a set $\{1, \ldots, n\}$ of consecutive integers.

(PC2) The modifier "physically" indicates that this performance is (in some manner) constrained by the limitations of physics (that is, of "the material, energetic world, epitomized by what physics is trying to talk about" spoken of by Campbell, 1988a, p. 440).

Then a "physically uncountable finite set" L is a set of "things" (of whatever sort, e.g., "NL expressions" or "atoms in Jupiter or the Solar System") such that the following propositions about L are known or assumed:

(PU1) L is Peano-finite in the sense of note 8.

(PU2) The limitations of physics make it impossible for any "physical computational system" to compute a one-to-one correspondence of L with a set $\{1, \ldots, n\}$ of consecutive integers, such as (PC1) and (PU1) together require.

I justify this interpretation chiefly by a passage from Luuk (2013, p. 90) that immediately precedes the previous passage quoted from that paper:

> The finity of NL comes from viewing language as a physical computational system rather than a purely mathematical object. As the spacetime that can support physical computational systems is finite (Krauss & Starkman, 2000), physical computational systems cannot, differently from mathematical objects, accommodate infinity.

Example: Language as a Humanly Uncountable Finity

Readers of this chapter who do not follow modern linguistics (by which I mean, not the body of research into language, but rather the behavior of the community of modern linguists) may find it odd that orthopractic linguists "view [...] language"—identified with "Natural Language"—as "a purely mathematical object," and odder yet that "[t]he finity of NL" could be in any doubt, let alone something that must be argued for (with little chance any time soon of being accepted by the linguistic community at large). A naïve outsider might even consider it obvious—except perhaps for the phrase "just physically uncountable"—that

> the maximum possible natural language "corpus"—everything that has ever been and will be processed—is not infinite but a finite, just physically uncountable set. We propose that this is precisely the nature of language as it should be accounted for [...]. (Luuk & Luuk, 2011, p. 7)

Obvious or not, from the quoted definition of "the maximum possible natural language 'corpus'" L together with standard assumptions about physics (having straightforward consequences for a "physical computational system" of any description), such as those about "spacetime" attributed to Krauss and Starkman (2000) by Luuk (2013) as well as some (there not mentioned) about the manner in which matter and energy inhabit "spacetime" (e.g., that there is a minimum spatial extent for any particle of matter, and a minimum quantum of energy), it is demonstrably true that L is a *subset* of some Peano-finite mathematical set of (nonmathematical) "things": the demonstration comes down to an arithmetic exercise in estimation. That L is itself Peano-finite—that is, that (PU1) holds for L— then follows from (iii) in note 8.

To prove that (PU2) also holds for L would require more detail than is given by Luuk and Luuk (2011) or Luuk (2013) as to what they would accept as a "physical computational system" and as to why no such system can, in the case of L, compute such a one-to-one correspondence as (PU2) requires. Rather than try to tease out their intended meaning here, I give a logically weaker notion of "humanly uncountable" that is (I hope) of quite wide applicability even though it is specially tailored to prove that the finite "natural language 'corpus'" L is a humanly "uncountable set."

Of course I define "humanly uncountable" to mean "cannot be humanly counted." I define the modified verb *humanly count* as follows.

(HC1) The core verb *count* is that particular kind of imagined terminating process described under the heading "Formation and use of mathematical finities: large or indefinite natural numbers" on p. 66.

(HC2) The modifier *humanly* indicates that the imagining of this process is (in some manner) constrained by limitations on the behavior of human subjects (including limitations that are simply of "the material, energetic world, epitomized by what physics is trying to talk about" and limitations that "also includ[e] the realities, if any, sociology"—and the other human sciences, biology, etc.—are "trying to describe," again following Campbell, 1988a, p. 440).

Then a "humanly uncountable finite set" L is a set of "things" (of whatever sort) in human experience, such that the following propositions about L are known or assumed.

(HU1) L is Peano-finite in the sense of note 8.

(HU2) Limitations on the behavior of human subjects make it impossible for any such subject to count L in the sense that (HC1) and (HU1) together require.

Note that in (HC2) and (HU2), "limitations on behavior" should be read broadly, to include limitations on perception, cognition, and so on.

That "naïve outsider" to modern linguistics, who I earlier imagined might find it odd that orthopractic linguists view "natural language" as a purely mathematical, mathematically infinite structure, might well also be astounded that Luuk and Luuk (2011, p. 7), having proposed that "natural language" is not a purely mathematical structure (but rather "everything" of a linguistic nature "that ever has been and will be processed") and that therefore "natural language" is not a mathematically infinite structure (but rather a mathematically "finite, just physically uncountable set"), should then have found it necessary in their very next sentence to call their proposal a "substantial correction [...] suggested for the sake of unambiguity and exactitude." Yet as a matter of fact, a central dogma of modern linguistics—with (as befits dogma) a long, if sparse, tradition of heretical dissent (e.g., Carroll, 1958; Dixon, 1963; Olmsted, 1967; Peters, 1972; Reich, 1969; see also Bickerton, 2009)—is that "natural language" is not only infinite, but it is *necessarily* infinite.

The argument for this dogma runs essentially as follows. First, equivocate between (1) "natural language" as the body of human language behavior (including imagined language behavior), that is, what Luuk and Luuk (2011) describe as "the maximum possible natural language 'corpus'" and I have denoted by *L*, and (2) "Natural Language" as (some sort of) a mathematical/ computational model of (some of) that behavior, which is often denoted by NL. Second, in order to account for structural features that are apparently absent from the communicative behavior (if any) of all non-human communities of living beings, but are—or are imagined to be—universally present in *L*, argue that NL must include some version of "recursion" (different ones have been proposed at different times). Third, argue (para)mathematically that the mathematical/computational model of (that version of) "recursion" necessarily generates a mathematical infinity of distinct outputs,[25] whence it follows that NL, *and therefore L*, is infinite.[26]

All three steps in this sequence are so naturalized in the linguistic community that to present a proof for the claim (in the quotation from an interview quoted on p. 55 under heading (T1a)) that "Merge is innate but that is almost automatic," Chomsky need only say "if there is a recursive system at all, if there is an infinite number of objects, you have to have Merge." This use of "if" is rhetorical as well as dialectical: he is not merely stating a logical proposition S of the form "If R, then M," a proposition that is logically agnostic as to the truth or falsity of R (since S is true not only when both R and M are true, but also when both R and M are false, and when R is false but M is true); he is stating S *in a particular kind of communicative context*, one in which *the truth of* R *is presupposed*, in order that he may apply *that presupposition together with* the logical structure of S to conclude the truth of M.

The Luuks' proposal has two parts: it defines L in such a way that it is manifestly finite, and therefore not the same as NL, which they accept is mathematically infinite because it admits recursion; and it affirms that L, though mathematically finite, is "physically uncountable," which (whatever it means precisely) is apparently sufficient to account for "the nature of language as it should be accounted for"—in particular, for those structural features that NL accounts for by recursion. They spell out this second part as follows (Luuk & Luuk, 2011, p. 7):

> 1. There is no data available on brain function that would implicate neurally implemented recursion. 2. It is not at all clear that an input procedure in the brain can be formally and straightforwardly derived from its final output with any certainty (much less by the laws of classical logic, as in the Chomskyan derivation of neurally implemented recursion). 3. The fact that natural language and arithmetic processing is by definition finite (although potentially physically uncountable) downplays the motivation for neurally implemented recursive process. 4. As recursion is primarily used for defining infinite sets, the only motivation for neurally implemented recursion is infinity in natural language and arithmetic competence.

They then (re)state what is, I think, their most important observation:

> [I]nfinity in natural language and arithmetic competence *reduces to imagining, not processing infinite embedding or concatenation.* A recursive process or procedure is not needed to account for this. In fact, there is no compelling evidence that the brain behaves in any recognizable way like the models of computation we use in formal logic. (Luuk & Luuk, 2011, p. 7, emphasis added)

In other words, the purported infinity of "Natural Language" NL has no impact on the finity or infinity of "natural language" L, because it is of the same sort as the purported infinity of **N**: NL exists (for some linguists and other human subjects) as a "single object" for "free use," but a commitment to its existence does not entail a mathematical infinity of further ontological commitments, one for each of the "infinite number of objects" in NL.

Like **N**, which I described on p. 70 as "formed in the act of imagining a *non-terminating process*—for **N**, the fundamental such process" (of passing from an integer to its successor), NL is formed in acts of "imagining, not processing" various more or less fundamental non-terminating processes of "embedding or concatenation," such as (as in Hauser, Chomsky, & Fitch, 2002, p. 1571, quoted by Luuk & Luuk, 2011, p. 9) making a new and longer sentence from "any candidate sentence [...] by [...] embedding it in 'Mary thinks that...'" or (as in Dixon, 1963, p. 83, quoted by Chomsky,

1963, p. 19) forming "sentences which each consist of a conjunction of clauses," of which "we are clearly unable to say that there is any definite number, N, such that no sentence contains more than N clauses." In the case of **N**, on p. 69 I wrote that

> I accept that for, at least, certain human subjects [...], **N** and similar mathematical structures comprising a mathematical infinity of substructures do "exist" in an "as if" way—that is, *"as if" they have been constructed by human subjects* within their "constructions of reality"

but also that

> I deny that this either implies or is implied by the claim that each of those substructures [...] "exists" for those human subjects, *even in an "as if" way.*

I think that in this latter way, too, NL is like **N**. That is, I deny that each of the supposed by-products of such imagined non-terminating process of "embedding or concatenation" (etc.) "exists" even in an "as if" way. I have no room for the argument in detail here, but it is very similar to the one embedded in my discussion of van Dantzig's "strict finitism" on pp. 70–72.

Finite Continua

In the half-century since its publication, Zeeman's "The Topology of the Brain and Visual Perception" (1962) has been frequently cited (examples relevant to aspects of the human sciences include Frank, 1992; Han, Humphreys, & Chen, 1999; and Minsky & Papert, 1967). It seems, however, that almost all these citations refer only to the theory of "tolerance spaces" introduced there. The warning—directed explicitly to physicists but implicitly to scientists in general—against the use of the real number system **R**, that

> any use of differential equations presupposes a 3-dimensional mathematical ether, which is every bit as pernicious as the physical ether of the last century. [...] The root of the trouble lies in the effort to fit the strait jacket of Euclidean space onto the universe. After all, nothing in physics suggests the existence of so sophisticated a mathematical construction as the real numbers [...]. Nothing in physics suggests even non-countability (Zeeman, 1962, pp. 255–256)

has been widely unacknowledged and widely unheeded (even by Zeeman, 1977, in his own later work on catastrophe theory, including applications to biology and the human sciences). In particular, human scientists of a quantitative disposition mostly continue to be enthusiasts for fitting "the strait jacket of Euclidean space onto the universe" via methodologies (in statistics, psychophysics, etc.—and catastrophe theory, see Gottman, Murray, Swanson, Tyson, & Swanson, 2003) that presuppose **R**, which as a set is not only mathematically infinite but *uncountably* infinite in the sense of Cantor (1874).

Notwithstanding my doubt (expressed on pp. 61–62) that an ontological commitment to the real numbers is "worth the while of the human sciences" (p. 62), it is clear to me that certain human subjects (myself included!) do make that commitment to **R**, which, like **N**, can be said to "exist" for us in an "as if" way for "free use" as a "single object" without admitting that any of us thereby make a further mathematical infinity of commitments, whether that be one commitment for each of the uncountably many "real numbers," or merely one for each of the countably infinite set of "'computable' numbers [that] may be described briefly as the real numbers whose expressions as a decimal are calculable by finite means" (Turing, 1936, p. 230). Just as in my account of the "formation" of **N**, I am happy to allow that (for us) **R** as a single object "is formed in the act of imagining a non-terminating process"; the process for **R** is more complicated than it is for **N**, and few outside the community of working mathematicians are likely ever to have become acquainted with it, but after two millennia of development from Eudoxus (c. 367 BCE) through Dedekind (1888) that process has now settled into the "benign fixation of mathematical practice." (For Turing's "computable numbers," the terminating process is more like that for **N**—both are, in fact, Turing machines—but still too complicated to describe in any detail here.)

Also as in my account of **N**, I am happy to allow that a human subject can have ontological commitments to some (humanly countable) finite number of individual members (or subsets, etc.) of **R**. Indeed, since **N** is commonly and not incorrectly understood to be a subset of **R**, most persons who have learned to count can be said already to have made some such commitments even if they have never heard of **R**.

However, all that being the case I remain skeptical that it is either prudent or necessary for human subjects "doing human science" (as contrasted with "doing mathematics" or, perhaps, "doing physics") to embrace **R**—even "in the context of continuous scaling and measurement" as in the example texts under (T1b), where I propose that better mathematical models (with the added virtue of not conflicting with von Uexküll's Axiom of Subjective Finiteness) can and should be constructed using *finite continua*.

Example: Finite Continua in Human Perception

For sound historical reasons, the polarity between *finite* (or, in some contexts, "countably infinite") and *infinite* (implicitly, uncountable) is often conflated with that between *discrete* and *continuous*. **R** is still sometimes called "the continuum"; the "Continuum Hypothesis," a famous problem in set theory (due to Cantor, 1878; made famous by Hilbert, 1900/1902; first solved in part by Gödel, 1940), is concerned with the possible "sizes" (in a certain sense) of subsets of **R**; from the proportions of Eudoxus and the geometry of Euclid, via calculus and differential equations (Newton, Leibniz, Euler, Laplace, etc.) to modern analysis, differential geometry, and analytic number theory (Dedekind and his contemporaries), mathematicians often found inspiration and motivation for what eventually would become the modern consensus definition of **R** in their efforts to model "continuous" phenomena geometrically, and **R** is now most often introduced to students as "the real number line."

Yet, in fact, for over 100 years, working mathematicians—though not the general scientific public, nor the public at large—have known that the two polarities are, logically, entirely separate: many mathematical structures that there are good reasons to call "discrete" are neither finite nor countably infinite; and, contrariwise, it is perfectly possible—and entirely natural in a mathematical sense—to ascribe "continuity" to an appropriate finite structure. What is more, I claim that finite continua are *also* natural in the sense that they can be good models of natural phenomena, specifically of phenomena in the human sciences, although they have not yet been used for that purpose as much as I would like to see them be.

Several types of mathematical structures on a finite set have been used to model continua in psychology.

- (Y, η), one of several tolerance spaces introduced by Zeeman, has as its underlying set "the right visual lobe, which we define to be the set of nerve cells in the lateral geniculate bodies that are stimulated by steady lights shining in X [the right visual field]. There are about a million cells in Y[...]" (Zeeman, 1962, p. 242). He defines "the set theory of a tolerance space" and proves that the set theory of (X, ξ)—the right visual field (assumed to be an infinite set of physical points on which a standard notion of distance, and therefore continuity, is supplied by the angular metric) with its perceptually defined "tolerance" relation—is isomorphic to that of (Y, η).
- E. Dzhafarov and D. Dzhafarov (2010a, 2010b, 2012) use the V-*spaces* introduced by Fréchet (1918) to good effect in their "behavioral" dissolution of the classical "Paradox of the Heap," which

  > can be stated as follows. (1) A single grain of sand does not form a heap, but many grains (say 1,000,000) do. (2) If one has a heap of sand, then it will remain a heap if one removes a single grain from it. (3) But, by removing from a heap of sand one grain at a time sufficiently many times, one can eventually be left with too few grains to form a heap. (E. Dzhafarov & D. Dzhafarov, 2012, pp. 105–106)

- I have proposed elsewhere (Rudolph, 2006c, p. 77) that finite simplicial complexes (FSCs) can provide "a mathematical formalization of the process of creating meaning out of ambivalence" (as described by Abbey & Valsiner, 2003), giving as examples:

- "full time," an FSC-based model for psychological time (Rudolph, 2006a)
- an FSC comprising eight nodes and eight edges that captures all the qualitative content of Russell's (1980) "circumplex model of affect" (Rudolph, 2006c, p. 77)
- a sequence of 19 FSCs—the first comprising two nodes and one edge, and the last 10 nodes, 23 edges, and 14 triangular faces—that, together, encode a (hypothetical) course of development of human "core color space" in accord with Berlin and Kay's (1969) data (Rudolph, 2006b, Fig. 5).

• More recently (Rudolph, 2008a, 2008b), I have undertaken the beginnings of a program to rehabilitate the "topological psychology" of Lewin (1936) and create an integrated mathematical model of his notion of "life space" and von Uexküll's (1920, 1926) notion of Umwelt, using *finite topological spaces* (more specifically, finite T_0 spaces; see Barmak & Minian, 2008).

Of all these, I am presently convinced that finite T_0 spaces may be the most useful in general, and certainly in the cases for which I previously proposed FSCs (of which they are a generalization). Figure 2.1 illustrates two examples. The first, in subfigures (a)–(c), is a 5-point T_0 space representing a simple linear scale with three "closed points" (the extreme values p and r and the intermediate value q) and two "open points" (the ambivalent—*not* intermediate!—values M and N): in (a), the space is presented entirely

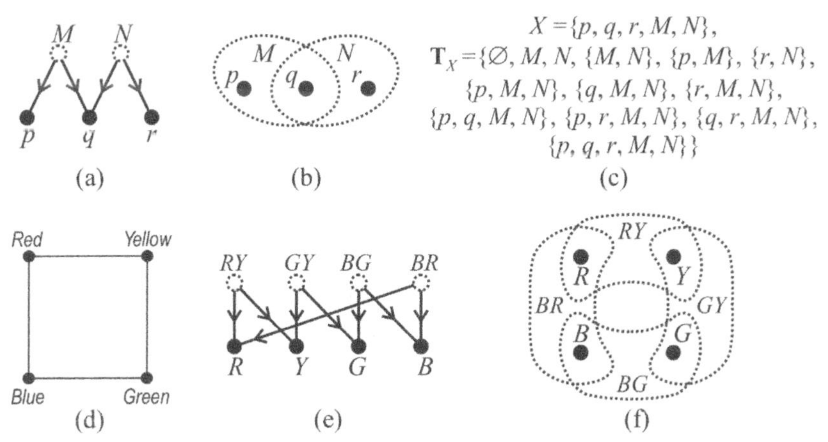

Figure 2.1

formally in *Hasse diagram* form (see Barmak & Minian, 2008, for details); in (b) it is depicted informally using a familiar and suggestive graphic convention; in (c) it is again presented entirely formally, but now in set-theoretic notation, by listing all the elements of the underlying set X of the space and of the topology $\mathbf{T}_X$ of the space (that is, the set of just those subsets of X that are to be considered "open"). The second example, in subfigures (d)–(f), illustrates the topologization of a simple model of human color-space: (d) is von Uexküll's first approximation to "a spatial form for the laws obtaining in the quality-circle of colours" (1926, p. 73), in essence a "circumplex" in the style of Russell (1980), with "four fixed points . . . of pure colour" indicated by dots (vertices), and between each appropriate pair "a straight line" (edge) indicating "the mixed colors formed from them" (Russell, 1980, pp. 73–74); (b) is a Hasse diagram derived from (a), in which I have taken the liberty of reconceptualizing von Uexküll's lines of "mixed colours" as ambivalences; and (c) is the informal graphic depiction of the T_0 space formally described by (b).

Unlike the (very simple) examples just described, finite T_0 spaces that can effectively model most continua arising "in the context of continuous scaling and measurement" (p. 55)—e.g., the "many 'physical phonetic' variables" mentioned by Sampson (1974, p. 236) in (T1b)—can be expected to have a reasonably large (but humanly countable) finity of points. To spell out the topology $\mathbf{T}$ of such a space is tedious and unenlightening, nor are informal graphics in the style of (b) and (f) very clear, either, when there are many points or more complicated ambivalences. In general, Hasse diagrams are much the clearest way to visualize finite T_0 spaces.

The Turtle That Stood on Its Own Back: Recursion Without Infinity

Near the beginning of the previous section I expressed my doubt that any uses of philosophically technical meanings of infinity in the human sciences, as in the example texts collected under the heading (T2), can fairly be said to entail infinitely many ontological commitments. Nonetheless, in at least some cases cited there (and many others not cited), some residual *horror infiniti*—usually associated to "recursion" in one of its many senses—clearly remains. Since I have made many references throughout this chapter to set theory and the axiomatic method, and have just now provided an example (albeit with few details) of how diagrammatic methods are used in rigorous mathematics (cf. Rudolph, 2012c), it seems reasonable to conclude with a proposal to let a diagram-based set-theoretic axiom banish all infinities from at least those senses of "recursion" that are in neither of the mathematical lineages, [α] and [β].

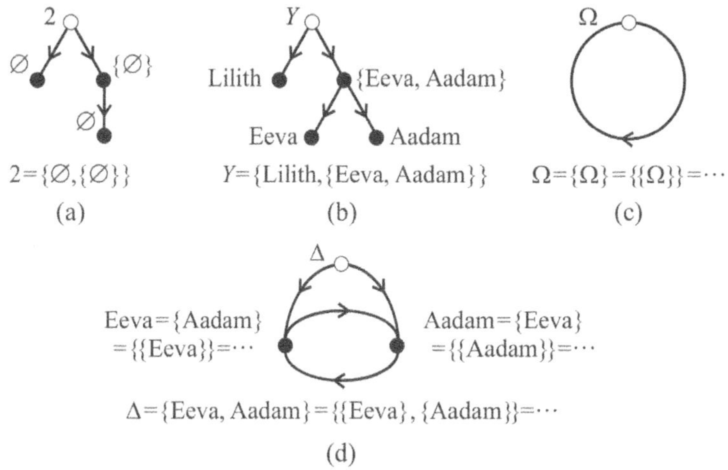

Figure 2.2

Figure 2.2 depicts various sets as "decorated accessible pointed graphs," a concept due to Aczel (1988), to which I refer the reader for the complete definitions and all the proofs (this monograph is surprisingly accessible, but for an alternative, equally friendly exposition with applications to semantics and philosophy, see also Barwise & Etchemendy, 1987); I include the figure simply to give some of the flavor of the theory. The definitions in brief are as follows: a *pointed graph* is a collection of *nodes* (denoted by small circles, filled except for one node in each graph, called its *point*) and *edges* that join pairs of nodes (denoted by lines or curves with arrows on them); such a graph is *accessible* if for each node there is a sequence of edges, with consistent arrows, from the point to that node, and *decorated* if to each node there is associated a set in such a way that the elements of that set are exactly the sets associated to the nodes (if any) joined to the given node by an edge whose arrow points away form the given node. Note that a "leaf node," one from which no arrow points away, can only be decorated with the empty set.

In (a), the "von Neumann ordinal number" 2—which in a familiar notation for finite sets is (as indicated) {∅,{∅}}, where ∅ is the equally familiar notation for the empty set—is depicted by a decorated accessible pointed graph. In (b), another set (of no special significance) is depicted by another accessible pointed graph: here the allowable decorations have been extended to include "atoms" and sets built from them, where

> [a]toms are objects that are not sets and are not made up of sets in any way, so that they have no set theoretical structure. But they can be used in the forma-

> tion of sets. See (Barwise 1975) for a discussion of the formalisation of set theory with atoms. In that book atoms are called Urelemente. (Aczel, 1988, p. 11)

The graphs in both (a) and (b) are *well-founded* in the sense (clearly very closely related to the sense of the term used in **R3**(b) on p. 26) that they contain no mathematically infinite path of consecutive edges (with consistently oriented arrows). Note that a mathematically infinite path of distinct consecutive edges would have to pass through a mathematical infinity of distinct nodes, but if edges are allowed to repeat—that is, if the path includes one or more loops—then there need only be finitely many nodes. In particular, the graphs in (c) and (d) are non-well-founded.

As a preliminary, Aczel shows that in standard set theory with the Axiom of Foundation, sets and well-founded decorated accessible pointed graphs are essentially equivalent; in particular, no non-well-founded accessible pointed graph can be decorated, and so such a graph does not depict any set. Similar results are true when atoms are allowed.

Next, Aczel considers a modification of standard set theory, in which the Axiom of Foundation is replaced by an "anti-foundation axiom" (AFA) stating, essentially, that every accessible pointed graph—whether or not it is well-founded—can be decorated so as to depict some set uniquely associated to that graph. In a tour de force, he proves that this modified set theory is equiconsistent with the usual one: If there are no contradictions in the usual theory, then there are none in the modified theory, and vice versa.

So, for instance, assuming AFA the simplest non-well-founded decorated accessible pointed graph, shown in (c), represents a unique set (which Aczel calls Ω) that is its own only element. Further, as Aczel (1988, p. 7) points out, although "the infinite expression" $\Omega = \{\Omega\} = \{\{\Omega\}\} = \ldots$ "associated with" this extraordinary set

> might suggest that in some sense Ω is an infinite object [...] a moment's thought should convince the reader that Ω is as finite an object as one could wish. After all it does have a finite picture. We may call sets that have finite pictures HEREDITARILY FINITE sets. (Aczel, 1988, p. 7)

Now, Ω is a purely set-theoretical set. But the hereditarily finite set Δ in (d) has atoms as decorations on its non-point nodes; interpreting those atoms as human subjects, we have in (d)—as it seems to me—a good first cut at a hereditarily finite mathematical model of several non-mathematical versions of "recursion" in the human sciences.

FINAL REMARKS AND ACKNOWLEDGMENTS

This chapter is in part a sequel to research performed (starting in 2004) with support from the National Science Foundation, via an Interdisciplinary

Grant in the Mathematical Sciences (DMS-0308894) and a Major Research Instrumentation Grant (BCS-0420939). It was catalyzed by the Kitchen Seminar at Clark University, convened by Jaan Valsiner and populated by an apparently endless series of interesting, interested participants. I dedicate it to Seymour Papert, an especially human scientist and talented recursor.

NOTES

1. Mathematicians working in "topological dynamics" use "recursive" for a technical notion (unique to that subfield) generalizing the more common technical notion (in dynamics) of "recurrent"; see Gottschalk and Hedlund (1955). This notion has nothing to do with (what I am calling) the "usual notion of 'recursion' in mathematics," nor has it had (to my knowledge) any influence at all on mathematics outside its subfield, much less in any science, so I will ignore it.
2. Define functions *f* and *g* for number inputs by these rules. (F) $f(1) = 1, f(2) = 1$, and $f(n) = 0$ for $n > 2$. (G) $g(n) = 1$ if there exist natural numbers x, y, and z such that $x^n + y^n = z^n$, and $g(n) = 0$ if there do not exist natural numbers x, y, and z such that $x^n + y^n = z^n$. Then *f* and *g* are in fact the same function; to prove this is to prove Fermat's Last Theorem (Wiles, 1995; Taylor & Wiles, 1995).
3. As a matter of fact, in the case of the Fibonacci number function *F*, the two definitions given are both qualitatively and quantitatively closer than they first appear. A naïve calculation of the *n*th power of any number x other than 0 and 1, in particular of the Golden Ratio φ, uses $n - 1$ recursively-structured repetitions of the single operation '×' of multiplication (namely, $x^2 = x^1 \times x$, $x^3 = x^2 \times x$, and so on to $x^n = x^{n-1} \times x$), and of course starts from the base case $x^1 = x$; so this particular definition by formula in fact is a lightly disguised definition by recursion. (Mathematicians have discovered procedures, and computer scientists have implemented them efficiently, to calculate the nth power of any given number in a fraction of $n - 1$ steps; but it can be proved that the best such procedures remain essentially recursive, and the number of steps still depends on n in a way that increases as n does.) On the other hand, many definitions by formula really do define functions that provably cannot be defined by any definition by induction. Unfortunately, I would have to go too far afield to supply an example of such a function (or such a proof).
4. It seems to me, as an outsider to recursion theory, that "benign fixation" has set in in favor of Soare's recommendations. One of the few (partial) dissenters is Arkoudis (2008). In any case, I have no need to take a position here on the recommendations.
5. I am deliberately ignoring the distinction between "total recursive functions" and "partial recursive functions," the latter being in fact the functions that can actually be "computed" by a Turing machine (see Soare, 1996, and references therein). This important distinction—as I understand it—captures a difference between "stopping" and "not stopping" that differs from the difference I am concerned with.

6. "Paramathematical discourse" and "metamathematics" are entirely distinct notions. Metamathematics, like geometry, topology, algebra, or analysis, is (just) a particular kind of mathematics, closely allied to mathematical logic; its practice consists in applying mathematical methods (specialized, but not peculiar) to mathematical proofs, definitions, and so on (Kleene, 1952). Paramathematical discourse is (just) discourse about a particular subject, namely, mathematics; its practice consists in talking and writing about mathematics. Of course, all "practice" of "applying mathematical methods" entails more or less talking and writing, so Kleene's book (for instance) is to that extent both metamathematical and paramathematical; but metamathematics makes up a negligible portion of all paramathematical discourse.
7. All these variants are attested in the sciences, social sciences, and mathematical sciences as represented in the Google Scholar and JSTOR corpora. Variants with *eternally* or *eternity* are much rarer there; they appear mostly in theological contexts, and seemingly (as of this writing) not at all in the sciences.
8. A set X is by definition: Dedekind-infinite if there is a subset Y of X, not identical to X, that is in one-to-one correspondence with X; Dedekind-finite if it is not Dedekind-infinite; Peano-finite if either X is empty (has no members) or there is a natural number n and a one-to-one correspondence of X with the set $\{1, \ldots, n\}$ of the first n natural numbers; Peano-infinite if it is not Peano-finite; denumerable (or countable) if and only if it is in one-to-one correspondence with a subset of the set $\mathbf{N}$ of natural numbers; and non-denumerable (or uncountable) if and only if it is not denumerable. In the standard set theory ZFC (i) a set X is Dedekind-finite if and only if X is Peano-finite, (ii) every (Peano- or Dedekind-) infinite set contains a denumerable infinite subset Y, and (iii) every subset of a (Peano- or Dedekind) finite set is finite (in the same sense). The proofs of (i)–(ii) depend on the Axiom of Choice, "C" in ZFC; without it, all three statements are false (see Potter, 2004).
9. Harries-Jones (1995, p. 186) quotes Bateson's account (as recorded in his notebook) of using the word "recursive" in speech "in a 1975 address to a meeting in London," and strongly suggests (ibid., p. 183ff.) that he had picked it up (from Varela) in 1973; but if Harries-Jones gives a pointer to an earlier printed instance, I have missed it.
10. Incidentally, although Scheff (2005, p. 384) writes of "representing a recursive model of frame and awareness structures with mathematical notation," his notation (which "follow[s] and extend[s]" Baptista, 2003, and is "almost identical to Lefebvre's (1977) [...] but slightly less complex") is not in fact "mathematical" as it stands: it is symbolic in a manner that certainly might facilitate, but does not by itself constitute, an application of mathematics to "frame and awareness structures." In particular it does not serve to connect example [24] and its like to any previously mathematicized notions of recursion.
11. Langendoen and Postal (1984) introduced non-denumerable (but still discrete) infinities into linguistics. Their work, of low mathematical quality, has been mathematically demolished repeatedly (e.g., by Bradfield, n.d.); it has never been retracted by its authors, nor built upon by others.

12. Circular reasoning is also related to "circular explanation" as defined and contrasted with "recursive explanation" by Lipton and Thompson (1988); see the brief quotation on pp. 50–51.
13. "Once, during my undergraduate years, I used the word 'infinity,' and my mathematics professor said, 'I won't have bad language in my class!'" (Courant, 1996, p. i).
14. I learned "finity" from Luuk and Luuk (2011). Although the word is somewhat out of fashion, it is not obsolete (see finity, n.d.); I think it's a grand word and would like to encourage its use.
15. Here and below I use "axiom" not in Euclid's sense of "a self-evident proposition" but in the sense of modern mathematics (already exemplified in the earlier discussion of axioms of set theory), and particularly of (mathematical) modeling: a proposition about the structure of some domain of inquiry assumed as a basis for further (mathematical) exploration of that domain.

    The situation has been nicely described by Jaan Valsiner (2009, p. 46) for readers who are in the business of "the observation of human systems": "axioms are taken as bases for knowing—following them proceeds without doubt. Yet choice of one's axioms is always deeply filled with doubt—this distinguishes the act of axiom construction from religious conversion." One is, of course, free to doubt some choices more than others.
16. Campbell (1974) gives an extensive historical review including a hundred or so bibliographical references; Cziko and Campbell (1990/1996) update those with several hundred more, and EE has continued to be very productive since 1996.

    Note that—like Lorenz and Campbell (but unlike a number of others whom Campbell judged to have contributed to "evolutionary epistemology" and who accepted that judgment, most notably Karl Popper)—I am not trained as a philosopher and do not purport to act professionally as a philosopher; neither am I nearly as well read in philosophy as Campbell's and Lorenz's works make it clear they were. It is likely, therefore, that some (I hope not all) of my speculative pronouncements on ontology are not merely naïve and/or amateurish, but also obviously refutable by sophisticated and/or professional philosophers, whose indulgence I beg and whose corrections I solicit. In the meantime, non-philosophically trained readers from the human sciences have been warned.
17. In general I prefer to use the word "things" for what are more formally called "objects," "existents," "referents of knowledge" (Campbell's choice), etc.; but for mathematics I prefer "structures" to "things," partly for rhetorical reasons, mostly because so many referents of specifically mathematical knowledge do seem (to me and, I think, to most working mathematicians) to be highly "structured"—including even such referents as natural numbers (see the brief discussion of "*ante rem* as opposed to *in re* structures" by Burgess, 1999, pp. 286–287, and more generally Burgess's survey in that book review of various notions of "structure" in the philosophy of mathematics—to none of which I am competent either to claim or deny allegiance in my own, at best psychologically inspired, uses of the term).

18. My distinguished friend John Burgess, who is eminent both as a mathematical logician and as a philosopher of mathematics and science, appears to differ from me here when he writes that

    > the avoidance of ontological commitments to abstract entities does not seem to have won recognition in the scientific community as being in itself a goal of the scientific enterprise on a par with scope and accuracy, and convenience and efficiency, in the prediction and control of experience. It seems, on the contrary, a matter to which most working scientists attach no importance whatsoever. It seems distinctively and exclusively a preoccupation of philosophers of a certain type.... One would search the physics journals in vain for any expression of nominalistic qualms and scruples, of reluctance and hesitancy to use mathematical apparatus, of suspicion that such "Platonistic claptrappery" as complex numbers may be a source of "confusion and self-deception." (Burgess, 1983, p. 98)

    I continue to stand (or hide) behind my disclaimers in note 16, but do note that (i) I am not at all calling for "the avoidance of ontological commitments to abstract entities," but merely (?) for the avoidance of ontological commitments to infinitely many entities, whether "abstract" or concrete, (ii) I am (grudgingly) open to allowing such commitments in "the physical sciences," just not in the human sciences, and (iii) one would not, in fact, search the psychology literature entirely in vain for an expression of such finitistic scruples (see, e.g., Rudolph, 2006a,b,c, 2008a,b; Valsiner & Rudolph, 2012).
19. Here and in the remainder of this chapter, "subject" and "subjective" (*Subjekt* and *subjektiv* in the original, von Uexküll, 1920, p. 57), are used with their standard meanings in German (as well as, e.g., Russian) psychology (at least in the early 20th century), which are close to their standard meanings in English grammar, but not at all like their standard meanings in (contemporary) English-language psychology. I am indebted to members of the Kitchen Seminar for reminding me that a note like this would probably be helpful to many of my readers, and especially to Irina Mironenko for the information about Russian psychology.
20. Rudolph (n.d.) sketches what might make such a generalization "suitable," and how the generalizations proposed by von Uexküll, 1933—summarized, with apparent justice, by Stjernfelt, 2007, p. 459, note 269, as "no less than a fascist biologist doctrine of state"—might (quite aside from political philosophy or ethics) not be "suitable."
21. It is worth pointing out that not only "ordinary" speech, but also both subvocal and imagined speech, can be paramathematical: such behaviors are common among mathematicians "doing mathematics," and not uncommon among persons in general when they calculate.
22. The number with the standard notation $10^{10^{100}}$ and the name "googolplex" (Kasner & Newman, 1940, p. 23) is far larger than either $10^{10^{10}}$ or 2^{65536}, named by D. van Dantzig (1955/1989) and D. Isles (1992) in the titles of their respective papers, each of which points out that the named number can not be physically counted off (or physically printed in base one or base 10).

23. I see little similarity (or agreement) between the ideas propounded by Lakoff and Núñez (2000) and those I am trying to formulate—in particular, between their idea of the Basic Metaphor of Infinity (op. cit., p. 155ff.) and my ideas on the status of mathematical infinities in human subjects' ontologies—despite our shared commitment to (possibly incompatible versions of) a notion of "embodied mind."
24. This distinction—between the kind of "as if" existence I ascribe to **N**, which does not entail any kind (even an "as if" kind) of existence for all its substructures (e.g., its members and subsets), and a more traditional kind of existence (that might be called "hereditary"), which would entail (and be entailed by) the (same kind of) existence for all substructures—seems similar to the classical distinction between "potential infinities" and "actual infinities." Perhaps it is identical, but I cannot assert that because I do not have a sufficiently clear understanding of the classical distinction.
25. Even if "physically uncountable" in the sense of Luuk and Luuk (2011) and Luuk (2013), such a mathematical infinity is countable in the set-theoretical sense (note 8), the claims of Langendoen and Postal (1984) notwithstanding (cf. note 11).
26. Paul Krugman describes a somewhat similar situation, of dogmatic attachment to 'elegant' mathematical models, in macroeconomics.

> The freshwater view began with an intellectually appealing solution in terms of rational behavior and imperfect information that unfortunately turned out to be all wrong; rather than drop their modeling strategy and the nifty math that went with it, freshwater economists decided to deny the facts instead. (Krugman, 2013)

REFERENCES

Abelson, H., Goodman, N., & Rudolph, L. (1974). *LOGO Manual* (Technical Report No. AIM-313). Cambridge, MA: Massachusetts Institute of Technology Artificial Intelligence Laboratory. Retrieved from http: //hdl.handle.net/1721.1/6226

Abbey, E., & Valsiner, J. (2004). Emergence of meanings through ambivalence. *FQS: Forum Qualitative Sozialforschung, 6*(1), Art. 23. Retrieved from http://www.qualitative-research.net/fqs-texte/1-05/05-1-23-e.htm

Aczel, P. (1988). *Non-well-founded sets.* Stanford, CA: Stanford University Center for the Study of Language and Information. Retrieved from http://standish.stanford.edu/pdf/00000056.pdf

Alexander, K. L., & Pallas, A. M. (1983). Bringing the arrows back in: On the recursivity assumptions in school process models. *Social Forces, 62*(1), 32–35.

Aposiopesis. (1885/2013). Oxford English dictionary (1st ed.). Oxford, England: Oxford University Press. Retrieved from http://www.oed.com/view/Entry/9392

Arkoudis, K. (2008). Computation, hypercomputation, and physical science. *Journal of Applied Logic, 6*, 461–475.

Aristotle. (1902). *Selecta ex Organo Aristoteleo: Capitula in usum scholarum academicarum* [Selections from the Organon of Aristotle: Chapters used in the University Schools] (J. R. Magrath, Ed.). Oxford, England: Clarendon Press.

Armstrong, D. M. (1973). *Belief, truth, and knowledge.* Cambridge, England: Cambridge University Press.

Azzouni, J. (2006). How and why mathematics is unique as a social practice. In R. Hersh (Ed.), *18 unconventional essays on the nature of mathematics* (pp. 201–219). New York, NY: Springer.

Baptista, L. C. (2003). Framing and cognition. In J. Trevino (Ed.), *Goffman's legacy* (pp. 197–215). Lanham, MD: Rowan & Littlefield.

Bar-Hillel, Y. (1954). Logical syntax and semantics. *Language, 30*(2), 230–237.

Barmak, J. A., & Minian, E. G. (2008). Simple homotopy types and finite spaces. *Advances in Mathematics, 218*, 87–104.

Bartlett, F. C. (1932). *Remembering: A study in experimental and social psychology.* Cambridge, England: Cambridge University Press.

Barwise, J. (1975). *Admissible sets and structures.* New York, NY: Springer.

Barwise, J., & Etchemendy, J. (1987). *The liar: An essay in truth and circularity.* New York, NY: Oxford University Press.

Bateson. G. (1955/1987). Epidemiology of a schizophrenia. In G. Bateson, *Steps to an ecology of mind: Collected essays in anthropology, psychiatry, evolution, and epistemology* (pp. 199–204). Northvale, NJ: Jason Aronson. (Originally published in A.P.A. Psychiatric Research Reports, 1955, II, 39–51)

Bateson. G. (1955/1987). A theory of play and fantasy. In G. Bateson, *Steps to an ecology of mind: Collected essays in anthropology, psychiatry, evolution, and epistemology* (pp. 183–198). Northvale, NJ: Jason Aronson Inc. (edited version of a talk given in May, 1955)

Bateson, G. (1960/1987a). The group dynamics of schizophrenia. In G. Bateson, *Steps to an ecology of mind: Collected essays in anthropology, psychiatry, evolution, and epistemology* (pp. 233–248). Northvale, NJ: Jason Aronson. (Originally published in L. Appleby, J. M. Schler, & J. Cumming (Eds.), *Chronic schizophrenia: Explorations in theory and treatment*, pp. 90–104. Glencoe, IL: The Free Press,1960)

Bateson, G. (1960/1987b). Minimal requirements for a theory of schizophrenia. In G. Bateson, *Steps to an ecology of mind: Collected essays in anthropology, psychiatry, evolution, and epistemology* (pp. 249–274). Northvale, NJ: Jason Aronson Inc. (Originally published in *Archives of General Psychiatry*, 1960, 2, 477–491)

Bateson, G. (1969/1987). Double bind, 1969. In G. Bateson, Steps to an ecology of mind: Collected essays in anthropology, psychiatry, evolution, and epistemology (pp. 276–283). Northvale, NJ: Jason Aronson. (Originally given as a paper in August, 1969, at a Symposium on the Double Bind)

Bateson, G. (1971/1987). The cybernetics of 'self': A theory of alcoholism. In G. Bateson, *Steps to an ecology of mind: Collected essays in anthropology, psychiatry, evolution, and epistemology* (pp. 315–344). Northvale, NJ: Jason Aronson. (Originally published in *Psychiatry*, 1971, 1, 1–18)

Bateson, G. (1977). Afterword. In J. Brockman (Ed.), *About Bateson: Essays on Gregory Bateson* (pp. 235–247). New York, NY: E. P. Dutton.

Bateson, G. (1980). *Mind and nature: A necessary unity.* New York, NY: Bantam Books.

Bateson, G. (1987). *Steps to an ecology of mind: Collected essays in anthropology, psychiatry, evolution, and epistemology.* Northvale, NJ: Jason Aronson.

Bateson, G., Jackson, D. D., Haley, J., & Weakland, J. H. (1956/1987). Toward a theory of schizophrenia. In G. Bateson (Ed.), *Steps to an ecology of mind: Collected essays in anthropology, psychiatry, evolution, and epistemology* (pp. 205–232). Northvale, NJ: Jason Aronson. (Originally published in *Behavioral Science,* 1956, 1(4), 251–264)

Berlin, B., & Kay, P. (1969). *Basic color terms: Their universality and evolution.* Berkeley, CA: University of California Press.

Bertalanffy, L. von. (1967). *Robots, men, and minds: Psychology in the modern world.* New York, NY: George Braziller.

Bickerton, D. (2009). Recursion: Core of complexity or artifact of analysis? In T. Givón & M. Shibatani (Eds.), *Syntactic complexity: Diachrony, acquisition, neuro-cognition, evolution* (pp. 531–544). Amsterdam, NL: John Benjamins.

Bonsall, F. F. (1982). A down-to-earth view of mathematics. *The American Mathematical Monthly, 89*(1), 8–15.

Bradfield, J. C. (n.d.). *The absurdity of vastness.* School of Informatics, University of Edinburgh, Edinburgh, UK. Retrieved from http://www.osti.gov/eprints/topicpages/documents/record/071/2499072.html

Brink, D. O. (1989). *Moral realism and the foundations of ethics.* Cambridge, England: Cambridge University Press.

Brouwer, L. E. J. (1907). *Over de grondslagen der wiskunde* [On the foundations of mathematics]. Amsterdam, NL: Maas & Van Suchtelen.

Bruner, J. (1986). *Actual minds, possible worlds.* Cambridge, MA: Harvard University Press.

Burgess, J. P. (1983). Why I am not a nominalist. *Notre Dame Journal of Formal Logic, 24*(1), 93–105.

Burgess, J. P. (1999). Untitled [Review of the book *Philosophy of mathematics: Structure and ontology*]. *Notre Dame Journal of Formal Logic, 40*(2), 283–291.

Burke, M. (2010). *Literary reading, cognition and emotion: An exploration of the oceanic mind.* New York, NY: Taylor & Francis.

Caley, M. T., & Sawada, D. (1986). Recursive complementarity. *Cybernetica, 29,* 26–275.

Campbell, D. T. (1974). Evolutionary epistemology. In P. A. Schilpp (Ed.), *The philosophy of Karl Popper* (Part 1, pp. 413–463). La Salle, IL: Open Court.

Campbell, D. T. (1988a). Descriptive epistemology: Psychological, sociological, and evolutionary. In E. S. Overman (Ed.), *Methodology and epistemology for social sciences: Selected papers* (pp. 435–486). Chicago, IL: University of Chicago Press. (From the William James Lectures of 1977, Harvard University)

Campbell, D. T. (1988b). Perspective on a scholarly career. In E. S. Overman (Ed.), *Methodology and epistemology for social sciences: Selected papers* (pp. 1–26). Chicago, IL: University of Chicago Press.

Cantor, G. (1874). Über eine Eigenschaft des Inbegriffes aller reellen algebraischen Zahlen [On a property of the set of all real algebraic numbers]. *Journal für die reine und angewandte Mathematik, 77,* 258–262.

Cantor, G. (1878). Ein Beitrag zur Mannigfaltigkeitslehre [A contribution to the theory of infinite sets], *Journal für die Reine und Angewandte Mathematik, 84*, 242–258.

Carroll, J. B. (1958). An operational model for language behavior. *Anthropological Linguistics, 1*(1), pp. 37–54.

Carnap, R. (1929). *Abriss der Logistik* [Outline of logic]. Vienna, AT: Springer.

Chomsky, N. (1955) Logical syntax and semantics: Their linguistic relevance. *Language, 31*(1), 36–45.

Chomsky, N. (1957). *Syntactic structures.* The Hague, NL: Mouton.

Chomsky, N. (1963). Topics in the theory of generative grammar. In T. A. Sebeok (Ed.), *Current trends in linguistics* (Volume 0, pp. 1–57). The Hague, NL: Mouton.

Chomsky, N. (2009). Comments and answers. In A. S. Özsoy & M. Nakipoğlu (Eds.), *Noam Chomsky on language and cognition* (pp. 22–44). Munich, Germany: Lincom Europa.

Church, A. (1937). Untitled [Review of the paper "On computable numbers, with an application to the Entscheidungsproblem"]. *The Journal of Symbolic Logic, 2*(1), 42–43.

Clarke, J. W. (1981). Emotional deprivation and political deviance: Some observations on Governor Wallace's would-be assassin, Arthur H. Bremer. *Political Psychology, 3*(1/2), 84–115.

Clifford, P. A., Friesen, S., & Jardine, D. (2001). *The ontology of hope: Lessons from a child.* Paper presented at AERA 2001: What We Know and How We Know It, the 82nd Annual Meeting of the American Educational Research Association, Seattle, WA (April 10–14). Retrieved from http://www.udel.edu/aeracc/papers/The%20Ontology%20of%20Hope%202001.html

Courant, E. D. (1996). Foreword. In Courant, R., & Robbins, H. (Eds.), *What is mathematics?* (2nd edition, revised by I. Stewart), New York, NY: Oxford University Press.

Cziko, G. A., & Campbell, D. T. (1990/1996). Comprehensive evolutionary epistemology bibliography. *Journal of Social and Biological Structures, 13*(1), 41–82.

Dedekind, R. (1888). *Was sind und was sollen die Zahlen?* [What are numbers and what should they be?] Braunschweig, Germany: Friedrich Vieweg und Sohn.

De Morgan, A. (1838). Induction (mathematics). In *Penny Cyclopaedia of the Society for the Diffusion of Useful Knowledge* (Vols. 11–12, pp. 465–466). London, England: Charles Knight & Co.

Dickinson, J. (2006, May 2). [Blog comment]. Retrieved from http://www.volokh.com/posts/1146550392.comments.html

Dieudonné, J. (1984). MR0686341 [Review of the article "Dedekind's invention of ideals"]. *Mathematical Reviews.* Retrieved from http://www.ams.org/mathscinet-getitem?mr=0686341

Diriwächter, R. (2012). Structure and hierarchies in Ganzheitspsychologie. In L. Rudolph (Ed.), *Qualitative mathematics for the social sciences: Mathematical models for research on cultural dynamics* (pp. 189–226), New York, NY: Routledge.

Dixon, R. M. W. (1963). *Linguistic science and logic.* The Hague, NL: Mouton.

Dobuzinskis, L. (1987). *The self-organizing polity: An epistemological analysis of political life.* Boulder, CO: Westview Press.

Doll, W. E. (1993). Curriculum possibilities in a "post"-future. *Journal of Curriculum and Supervision, 8*(4), 277–292.

Doran, C. (1989). Jumping frames: Reflexivity and recursion in the sociology of science. *Social Studies of Science, 19*(3), 515–531.

Dzhafarov, E. N., & Dzhafarov, D. D. (2010a). Sorites without vagueness I: Classificatory sorites. *Theoria, 76*(1), 4–24.

Dzhafarov, E. N., & Dzhafarov, D. D. (2010b). Sorites without vagueness II: Comparative sorites. *Theoria, 76*(1), 25–53.

Dzhafarov, E. N., & Dzhafarov, D. D. (2012). The sorites paradox: A behavioral approach. In L. Rudolph (Ed.), *Qualitative mathematics for the social sciences: Mathematical models for research on cultural dynamics* (pp. 105–136), New York, NY: Routledge.

ellipsis. (1891/2013). Oxford English dictionary (1st ed.). Oxford, England: Oxford University Press. Retrieved August 8, 2013, from http://www.oed.com/view/Entry/60527

Ésénine-Volpine, A. S. Y. (1961). Le programme ultra-intuitionniste des fondements des mathématiques [The ultra-intuitionist program for foundations of mathematics]. In *Infinitistic Methods: Proceedings of the Symposium on the Foundations of Mathematics, Warsaw, 1959* (pp. 201–223). Oxford, England: Pergamon.

Eyde, D. B. (1983). Recursive dualism in the Admiralty Islands. *Journal de la Société des Océanistes, 39*, 3–12.

Finity. (1896/2013). Oxford English dictionary (1st ed.). Oxford, England: Oxford University Press. Retrieved from http://www.oed.com/ view/Entry/70468

Flavell, J. H. (1963). *The developmental psychology of Jean Piaget.* New York, NY: Van Nostrand.

Flavell, J. H. (1996). Piaget's legacy. *Psychological Science, 7*(4), 200–203.

Flavell, J. H., Botkin, P. T., Fry, C. L., Wright, J. W., & Jarvis, P. E. (1968). *The development of role taking and communication skills in children.* New York, NY: Wiley.

Flood, M. M. (1935). Division by non-singular matric polynomials. *Annals of Mathematics, 36*(4), 859–869.

Flood, M. M. (1938). Recursive methods and the analysis of time series. In *Report of fourth annual research conference on economics and statistics, Colorado Springs, 1938* (pp. 90–92). Colorado Springs, CO: Cowles Commission for Research in Economics.

Fornell, C. (1984). *A second generation of multivariate analysis: Classification of methods and implications for marketing research.* Working paper no. 414, Graduate School of Business Administration, University of Michigan.

Frank, A. U. (1992). Qualitative spatial reasoning about distances and directions in geographic space. *Journal of Visual Languages and Computing, 3*(4), 343–371.

Fréchet, M. (1918). Sur la notion de voisinage dans les ensembles abstrait [On the notion of neighborhood in abstract sets]. *Bulletin des Sciences Mathématiques, 42*, 138–156.

Frisch, R. (1933). Propagation problems and impulse problems in dynamic economics. In K. Kock (Ed.), *Economic essays in honour of Gustav Cassel* (pp. 171–205). London, England: Allen & Unwin.

Gandy, R. O. (1988). The confluence of ideas in 1936. In R. Herken (Ed.), *The universal Turing machine: A half-century survey* (pp. 55–111). Oxford, England: Oxford University Press.

Giddens, A. (1979). *Central problems in social theory: Action, structure, and contradiction in social analysis.* London, England: Macmillan.

Gödel, K. (1934). On undecidable propositions of formal mathematical systems, Notes by S. C. Kleene and J. B. Rosser on lectures at the Institute for Advanced Study, Princeton, New Jersey, 1934, 30 pp.

Gödel, K. (1940). *The consistency of the axiom of choice and of the generalized continuum hypothesis with the axioms of set theory.* Princeton, NJ: Princeton University Press.

Goffman, E. (1963). *Stigma: Notes on the management of spoiled identity.* Englewood Cliffs, NJ: Prentice-Hall.

Goffman, E. (1969). *Strategic interaction.* Philadelphia, PA: University of Pennsylvania Press.

Goffman, E. (1974). *Frame analysis: An essay on the organization of experience.* Cambridge, MA: Harvard University Press.

Gottman, J. M., Murray, J. D., Swanson, C. C., Tyson, R., & Swanson, K. R. (2003). *The mathematics of marriage: Dynamics, non-linear models.* Cambridge, MA: MIT Press.

Gottschalk, W. H. & Hedlund, G. A. (1955). *Topological dynamics.* Providence, RI: American Mathematical Society.

Grassmann, H. (1861). *Lehrbuch der Mathematik für höhere Lehranstalten* [Textbook of mathematics for higher educational institutions]. Berlin, Germany: Enslin.

Greenberg, J. H. (1949). The logical analysis of kinship. *Philosophy of Science, 16*(1), 58–64.

Grice, H. P. (1989). *Studies in the way of words.* Cambridge, MA: Harvard University Press.

Grize, J.-B. (1960). Du groupement au nombre: essai de formalisation [From "groupement" to number: an attempt at formalization]. In P. Greco, J.-B. Grize, S. Papert, & J. Piaget (Eds.), *Problèmes de la cons[t]ruction du nombre. Études d'épistémologie génétique* (Vol. XI, pp. 69–96). Paris, France: Presses Universitaires de France.

Han, S., Humphreys, G. W., & Chen, L. (1999). Uniform connectedness and classical Gestalt principles of perceptual grouping. *Perception & Psychophysics, 61*(4), 661–674.

Harries-Jones, P. (1995). *A recursive vision: Ecological understanding and Gregory Bateson.* Toronto, CA: University of Toronto Press.

Harvey, J. H. (2000). *Give sorrow words: Perspectives on loss and trauma.* London, England: Psychology Press.

Hauser, M. D., Chomsky, N., & Fitch, W. T. (2002). The faculty of language: What is it, who has it, and how did it evolve? *Science, 298*(5598), 1569–1579.

Heath, T. L. (1908). *The thirteen books of Euclid's Elements* (Vols. 1–3). Cambridge, England: Cambridge University Press.

Heims, S. P. (1977). Gregory Bateson and the mathematicians: From interdisciplinary interaction to societal functions. *Journal of the History of the Behavioral Sciences, 13,* 141–159.

Heims, S. P. (1991). *The cybernetics group.* Cambridge, MA: MIT Press.

Herbrand, J. (1971). Unsigned note on Herbrand's thesis, written by Herbrand himself. In W. D. Goldfarb (Ed.), *Logical writings* (pp. 272–276). Dordrecht, NL: D. Reidel. (Original work published 1931)

Hilbert, D. (1900/1902). Mathematical problems. Lecture delivered before the International Congress of Mathematicians at Paris in 1900 (M. Winston, Trans.). *Bulletin of the American Mathematical Society, 8*, 437–479.

Hilbert, D. (1926). Über das Unendliche [On the infinite]. *Mathematische Annalen, 95*, 161–190.

Hofstadter, D. (1979). *Gödel, Escher, Bach: An eternal, golden braid.* New York, NY: Basic Books.

Hull, D. L., Langman, R. E., & Glenn, S. S. (2001). A general account of selection: Biology, immunology, and behavior. *Behavioral and Brain Science, 24*, 511–573.

Huntington, E. V. (1905). The continuum as a type of order: An exposition of the modern theory. *Annals of Mathematics, 6*(4), 151–184.

Infinite. (1900/2013). Oxford English dictionary (1st ed.). Oxford, England: Oxford University Press. Retrieved from http://www.oed.com/ view/Entry/95411

Infinity. (1900/2013). Oxford English dictionary (1st ed.). Oxford, England: Oxford University Press. Retrieved from http://www.oed.com/view/Entry/95429

Isles, D. (1992). What evidence is there that 2^{65536} is a natural number? *Notre Dame Journal of Formal Logic, 33*(4), 465–480.

James, W. (1908). *The varieties of religious experience: A study in human nature.* London, England: Longmans, Green and Co.

Joyce, D. (1998). Euclid's *Elements.* Retrieved from http://aleph0.clarku. edu/ ~djoyce/java/elements/

Kasner, E., & Newman, J. (1940). *Mathematics and the imagination.* New York, NY: Simon & Schuster.

Keynes, J. M. (1972). The end of laissez-faire. In E. Johnson & D. Moggridge (Eds.), *The collected writings of John Maynard Keynes* (Vol. 9: Essays in Persuasion, pp. 272–294). London, England: Macmillan. (Original work published 1927)

Kinsella, A. R. (2010). Was recursion the key step in the evolution of the human language faculty? In H. van der Hulst (Ed.), *Recursion and human language* (Studies in Generative Grammar, Vol. 104, pp. 179–191). New York, NY: Walter de Gruyter.

Kleene, S. C. (1938). On notation for ordinal numbers. *Journal of Symbolic Logic, 3*, 150–155.

Kleene, S. C. (1952). *Introduction to metamathematics.* Amsterdam, NL: North-Holland.

Kleene, S. C. (1981). Origins of recursive function theory. *Annals of the History of Computing, 3*(1), 52–67.

Krauss, L. M., & Starkman, G. D. (2000). Life, the universe, and nothing: life and death in an ever-expanding universe. *The Astrophysical Journal, 531*, 22–30.

Kreisel, G. (1978). Untitled [Review of the book *Wittgenstein's Lectures on the Foundations of Mathematics,* Cambridge, 1939]. *Bulletin of the American Mathematical Society, 84*, 79–90.

Krugman, P. (2013, December 22). Uber and the Macro Wars [blog post]. Retrieved from http://krugman.blogs.nytimes.com/ 2013/12/21/ uber-and-the-macro-wars/

Laing, R. D., Phillipson, H., & Lee, A. (1966). *Interpersonal perception.* New York, NY: Springer.

Laird, J. D. (2007). *Feelings: The perception of self.* New York, NY: Oxford University Press.

Lakoff, G., & Núñez, R. (2000). *Where mathematics comes from: How the embodied mind brings mathematics into being.* New York, NY: Basic Books.

Langendoen, D. T., & Postal, P. M. (1984). *The vastness of natural languages.* Oxford, England: Basil Blackwell.

Lefebvre, V. (1994). On self-reflexive and self-organizing systems. *Applied Ergonomics 1*, 6–9. (Original work published in Russian in 1965)

Lefebvre, V. (1977). *The structure of awareness: Toward a symbolic language of human reflexion* (A. Rapoport, Trans.). Newbury Park, CA: Sage.

Le Moigne, J.-L., & Morin, E. (1999). *L'intelligence de la complexité: cognition et formation* [The intelligence of complexity: cognition and training]. Paris, France: Edition L'Harmattan.

Levi ben Gershom. (1909). *Sefer maassei choscheb: Die Praxis des Rechner, ein hebräisch-arithmetisches Werk des Levi ben Gerschom aus dem Jahre 1321, zum ersten male herausgegeben und ins deutsche übertragen von Gerson Lange* [Sefer maassei choscheb: The practice of the Calculator, a Hebrew-Arithmetic work of Levi ben Gershom from the year 1321, for the first time published and translated into German by Gerson Lange]. Frankfurt am Main, Germany: Golde. (Original work published 1321)

Lewin, K. (1936). *Principles of topological psychology* (F. Heider & G. M. Heider, Trans.). New York, NY: McGraw-Hill.

Lipton, P. (1991). *Inference to the best explanation.* London, England: Routledge.

Lipton, P., & Thompson, N. S. (1988). Comparative psychology and the recursive structure of filter explanations. *The International Journal of Comparative Psychology, 1*(4), 215–229.

Lorenz, K. (1962). Kant's doctrine of the a priori in the light of contemporary biology (C. Ghurye, Trans.). In L. von Bertalanffy & A. Rapoport (Eds.), *General systems: Yearbook of the Society for General Systems Research* (Vol. VII, pp. 23–35). Ann Arbor, MI: Society for General Systems Research. (Original work published in German in 1941)

Luuk, E. (2013). The structure and evolution of symbol. *New Ideas in Psychology, 31*, 87–97.

Luuk, E., & Luuk, H. (2011). The redundancy of recursion and infinity for natural language. *Cognitive Processing, 12*(1), 1–11.

Mally, E. (1912). *Gegenstandstheoretische Grundlagen der Logik und Logistik* [Object-theoretic foundations of logic and reasoning]. Leipzig, Germany: J. A. Barth.

Marshall, D. (2008). The dangers of purity: On the incompatibility of "pure sociology" and science. *The Sociological Quarterly, 49*, 209–235.

Maturana, H. R. (1970). *Biology of cognition.* Biological Computer Laboratory Research Report BCL 9.0. Urbana IL: University of Illinois.

Maturana, H. R., & Varela, F. J. (1980). *Autopoiesis and cognition: The realization of the living* (pp. 5–58). Boston, MA: D. Reidel.

Maurin, A. (2007). Infinite regress: Virtue or vice? In T. Rønnow-Rasmussen, B. Petersson, J. Josefsson, & D. Egonsson (Eds.), *Hommage à Wlodek. Philosophical papers dedicated to Wlodek Rabinowicz* (not paginated). Lund, SE: Lund University. Retrieved from http://www. fil.lu.se/hommageawlodek/site/papper/MaurinAnnaSofia.pdf

Meinong, A. (1904). Über Gegenstandstheorie (On object theory). In A. Meinong (Ed.), *Untersuchungen zur Gegenstandstheorie und Psychologie* [Investigations in theory of objects and psychology] (pp. 1–51.) Leipzig, Germany: J. A. Barth.

Mendell, H. (2004). Supplement to Aristotle and mathematics [third of four]. In E. N. Zalta (Ed.), *The Stanford encyclopedia of philosophy* (Winter 2008). Retrieved from http://plato.stanford.edu/entries/aristotle-mathematics/supplement3.html

Metaphysics Research Lab. (2013). *Stanford encyclopedia of philosophy.* Stanford, CA: Stanford University. Available at http://plato.stanford.edu/

Miller, P. H., Kessel, F. S., & Flavell, J. H. (1970). Thinking about people thinking about people thinking about...: A study of social cognitive development. *Child Development, 41*(3), 613–623.

Minsky, M., & Papert, S. (1967). Linearly unrecognizable patterns. In J. T. Schwartz (Ed.), *Mathematical aspects of computer science: Proceedings of symposia in applied mathematics* (XIX, pp. 176–217). Providence, RI: American Mathematical Society.

Mirimanoff, D. (1917). Les antinomies de Russell et de Burali-Forti et le problème fondamental de la théorie des ensembles [The antinomies of Russell and Burali-Forti, and the fundamental problem of set theory]. *L'Enseignement Mathématique, 19,* 37–52.

Morin, E. (1973). *Le paradigme perdu: La nature humaine* [The lost paradigm: Human nature]. Paris, France: de Seuil.

Morin, E. (1977). *La nature de la nature* [The nature of nature]. Paris, France: de Seuil.

Münsterberg, H. (1900). *Grundzüge der Psychologie* [Handbook of psychology]. Leipzig, Germany: J. A. Barth.

Nelson, E. (1995). Ramified recursion and intuitionism. In A. Fruchard & A. Troesch (Eds.), *Colloque trajectorien à la mémoire de Georges Reeb et Jean-Louis Callot* (Prépublication IRMA). Strasbourg, France: Institut de Recherche Mathématique Avancée.

Nelson, E. (2011). Warning signs of a possible collapse of contemporary mathematics. In M. Heller & W. H. Woodin (Eds.), *Infinity: New research frontiers* (pp. 76–85). Cambridge, England: Cambridge University Press.

Olmsted, D. L. (1967). On some axioms about sentence length. *Language, 43*(1), 303–305.

Owen, O. F. (1853). *The Organon, or logical treatises, of Aristotle* (Vol. I). London, England: Bohn.

Palmer, K. (1996). *The future of software process.* Retrieved from http://archonic.net/wildsoft/wilds14b.pdf

Papert, S. (1960). Problèmes épistémologiques et génétiques de la récurrence [Epistemological and genetic problems of recursion]. In P. Gréco, J.-B. Grize, S. Papert, & J. Piaget (Eds.), Problèmes de la cons[t]ruction du nombre. Études d'épistémologie Génétique (Vol. XI, pp. 117–148). Paris, France: Presses Universitaires de France.

Papert, S. (1980). *Mindstorms: Children, computers, and powerful ideas.* New York, NY: Basic Books.

Pascual-Leone, J. (1997) Metasubjective processes: The missing lingua franca of cognitive science. In D. M. Johnson & C. E. Erneling (Eds.), *The future of the cognitive revolution* (pp. 75–101). New York, NY: Oxford University Press.

Peano, G. (1889). *Arithmetices principia, nova methodo exposita* [The principles of arithmetic, presented by a new method]. Turin, IT: Bocca.

Peano, G. (1891). Sul concetto di numero [On the concept of number]. *Rivista di Matematica, 1,* 87–102, 256–267.

Peters, C. R. (1972). Evolution of the capacity for language: A new start on an old problem. *Man, 7*(1), 33–49.

Piaget, J. (1942). *Classes, relations et nombres. Essai sur les groupement de logistique et sur la réversibilité de la pensée* [Classes, relations, and numbers. Essay on logical "groupement" and on the reversibility of thought]. Paris, France: J. Vrin.

Piaget, J. (1949). *Traité de logique. Essai de logistique opératoire* [Treatise on logic. Essay on operational logic]. Paris, France: Colin.

Piaget, J., & Inhelder, B. (1967). *A child's conception of space.* New York, NY: W. W. Norton.

Poddiakov, A. & Valsiner, J. (2012). Intransitivity cycles and their transformations: How dynamically adapting systems function? In L. Rudolph (Ed.), *Qualitative mathematics for the social sciences: Mathematical models for research on cultural dynamics* (pp. 343–391). New York, NY: Routledge.

Poincaré, H. (1902). *La science et l'hypothèse* [Science and hypothesis]. Paris, France: Flammarion.

Potter, M. (2004). *Set theory and its philosophy.* Oxford, England: Oxford University Press.

Quine, W. V. O. (1969). *Set theory and its logic* (rev. ed.). Cambridge, MA: Harvard University Press.

Quirk, R., Greenbaum, S., Leech, G., & Svartvik, J. (1985). *A comprehensive grammar of the English language.* New York, NY: Longman.

Rabinovitch, N. L. (1970). Rabbi Levi Ben Gershon and the origins of mathematical induction. *Archive for History of Exact Sciences, 6*(3), 237–248.

Rapoport, A. (2000). *Certainties and doubts: A philosophy of life.* Montreal, CA: Black Rose Books.

Reich, P. A. (1969). The finiteness of natural language. *Language, 45*(4), 831–843.

Reid, C. (1970). *Hilbert.* New York, NY: Springer.

Richmond, A. H. (1993). Reactive migration: Sociological perspectives on refugee movements. *Journal of Refugee Studies, 6*(1), 7–24.

Robben, A. C. G. M. (1989). Habits of the home: Spatial hegemony and the structuration of house and society in Brazil. *American Anthropologist, 91*(3), 570–588.

Rottleuthner, H. (1988). Biological metaphors in legal thought. In G. Teubner (Ed.), *Autopoietic law: A new approach to law and society* (pp. 97–127), New York, NY: Walter de Gruyter.

Rudolph, L. (2006a). The fullness of time. *Culture & Psychology, 12*(2), 157–186.

Rudolph, L. (2006b). Mathematics, models and metaphors. *Culture and Psychology, 12*(2), 245–265.

Rudolph, L. (2006c). Spaces of ambivalence: Qualitative mathematics in the modeling of complex fluid phenomena. *Estudios de Psicología, 27*(1), 67–83.

Rudolph, L. (2008a). A unified topological approach to Umwelts and life spaces. Part I: Umwelts and finite topological spaces. In R. I. Sokol-Chang (Ed.), *Relating to environments: A new look at Umwelt* (pp. 185–205). Charlotte, NC: Information Age.

Rudolph, L. (2008b). A unified topological approach to Umwelts and life spaces. Part II: Constructing life spaces from an Umwelt. In J. Clegg (Ed.), *The observation of human systems* (pp. 113–179). New Brunswick, NJ: Transaction Publishers.

Rudolph, L. (2012a). Logic in modeling, logics as models. In L. Rudolph (Ed.), *Qualitative mathematics for the social sciences: Mathematical models for research on cultural dynamics* (pp. 38–74), New York, NY: Routledge.

Rudolph, L. (2012b). The varieties of dynamic(al) experience. In L. Rudolph (Ed.), *Qualitative mathematics for the social sciences: Mathematical models for research on cultural dynamics* (pp. 279–319), New York, NY: Routledge.

Rudolph, L. (n.d.). *Funktionskreis as hermeneutische Zirkel: A study in evolutionary ontology*. Manuscript submitted for publication.

Rudolph, L., & Valsiner, J. (2012). Introduction: Mathematical models and social representation. In L. Rudolph (Ed.), *Qualitative mathematics for the social sciences: Mathematical models for research on cultural dynamics* (pp. 1–36), New York, NY: Routledge.

Russell, J. A. (1980). A circumplex model of affect. *Journal of Personality and Social Psychology, 39*(1, supp. 6), 1161–1178.

Sampson, G. (1974). Is there a universal phonetic alphabet? *Language, 50*(2), 236–259.

Sarbin, T. R. (2004) The role of imagination in narrative construction. In C. Daiute and C. Lightfoot (Eds.), *Narrative analysis: Studying the development of individuals in society* (pp. 5–20), Thousand Oaks, CA: Sage Publications.

Sawada, D., & Caley, M. T. (1986). Recursive complementarity in the cybernetics of education. *Cybernetica, 29*(2), 93–104.

Sawada, D., & Caley, M. T. (1993). Complementarity: A recursive revision appropriate to human science. *Anthropology of Consciousness, 4*(2), 1–8.

Scheff, T. (1967). Toward a sociological model of consensus. *American Sociological Review, 32*, 32–46.

Scheff, T. (2005). The structure of context: Deciphering Frame Analysis. *Sociological Theory, 23*(4), 368–385.

Simpson-Vlach, R., & Ellis, N. C. (2010). An academic formulas list: New methods in phraseology. *Applied Linguistics, 31*(4), 487–512.

Soare, R. I. (1996). Computability and recursion. *Bulletin of Symbolic Logic, 2*, 284–321.

Spellman, B. A., & Mandel, D. R. (1999). When possibility informs reality: Counterfactual thinking as a cue to causality. *Current Directions in Psychological Science, 8*(4), pp. 120–123.

Stein, M. (2011). Faith and the practising analyst. *Journal of Analytical Psychology, 56*(3), 397–406.

Stjernfelt, F. (2007). *Diagrammatology: An investigation on the borderlines of phenomenology, ontology, and semiotics.* New York, NY: Springer.

Taylor, R., & Wiles, A. (1995). Ring theoretic properties of certain Hecke algebras. *Annals of Mathematics, 141*(3), 553–572.

Tinbergen, J. (1935). Annual survey: Suggestions on quantitative business cycle theory. *Econometrica, 3*(3), 241–308.

Tjon Sie Fat, F. E. (1981). More complex formulæ of generalized exchange. *Current Anthropology, 22*(4), 377–390.

Toch, H. (1965). *The social psychology of social movements.* Indianapolis, IN: Bobbs-Merrill.

Towne, M. (2010). Usage of recursive in political science. *Political Science & Politics, 43*(2), 259–261.

Turing, A. M. (1936). On computable numbers, with an application to the Entscheidungsproblem. *Proceedings of the London Mathematical Society, 42*, 230–265.

Turner, B. S. (1999). *Classical sociology.* Thousand Oaks, CA: Sage.

Uexküll, J. von. (1920). *Theoretische biologie* [Theoretical biology]. Berlin, Germany: G. Paetel.

Uexküll, J. von. (1926). *Theoretical biology* (D. L. MacKinnon, Trans.). New York, NY: Harcourt, Brace & Company. (Translation of von Uexküll, 1920)

Uexküll, J. von. (1933). *Staatsbiologie* [Biology of the state] (2nd ed.). Hamburg, Germany: Hanseatische Verlagsanstalt. Retrieved from http://babel.hathitrust.org/cgi /pt?id=njp.32101068558020

Valsiner, J. (2009). Baldwin's quest: A universal logic of development. In J. Clegg (Ed.), *The observation of human systems* (pp. 45–82). New Brunswick, NJ: Transaction Publishers.

Valsiner, J., & Rudolph, L. (2012). Who shall survive? Psychology that replaces quantification with qualitative mathematics. In E. Abbey & S. Surgan (Eds.), *Emerging methods in psychology* (pp. 121–140). New Brunswick, NJ: Transaction Publishers.

Van Bendegem, J. P. (2010). Finitism in geometry. In E. N. Zalta (Ed.), *The Stanford Encyclopedia of Philosophy* (Summer 2013). Retrieved from http://plato.stanford.edu/entries/geometry-finitism/

Van Bendegem, J. P. (2012). A defense of strict finitism (B. Van Kerkhove & J. P. Van Bendegem, Trans.). *Constructivist Foundations, 7*(2), 141–149.

van Dantzig, D. (1989). Is $10^{10^{10}}$ a finite number? In R. L. Epstein & W. A. Carnielli (Eds.), *Computability: Computable functions, logic, and the foundations of mathematics* (pp. 258–261). Belmont, CA: Wadsworth & Brooks. (Original work published 1955)

Vitale, B. (1989). Elusive recursion: A trip in recursive land. *New Ideas in Psychology, 7*(3), 253–276.

von Foerster, H. (Ed.). (1952). *Cybernetics: Circular causal and feedback mechanisms in biological and social sciences.* New York, NY: Josiah Macy, Jr. Foundation.

von Neumann, J. (1925). Eine Axiomatiserung der Mengenlehre [An axiomatization of set theory]. *Journal für die reine und angewandte Mathematik, 154*, 219–240.

von Neumann, J. (1951). The general and logical theory of automata. In L. A. Jeffress (Ed.), *Cerebral mechanisms in behavior: The Hixon symposium* (pp. 1–31), New York, NY: Wiley.

Walmsley, D. J. (1971). *A systems approach to consumer spatial behaviour.* Unpublished doctoral dissertation, The Australian National University, Canberra, AU.

Walmsley, D. J. (1972). *Systems theory: A framework for human geographical enquiry.* Department of Human Geography Publication HG/7. Canberra, AU: The Australian National University.

Walmsley, D. J., & Day, R. A. (1972). *Perception and man–environment interaction: A bibliography and guide to the literature.* Armidale, AU: Geographical Society of New South Wales, New England Branch. (Issue 2 of Occasional Papers in Geography)

Walmsley, D. J. (1973). The simple behaviour system: An appraisal and an elaboration. Geografiska Annaler. Series B, *Human Geography, 55*(1), 49–56.

Walmsley, D. J., & Lewis, G. J. (1984). *Human geography: Behavioural approaches.* Harlow, England: Longman.

Werner, H. (1959). *Einführung in die Entwicklungspsychologie* [Introduction to developmental psychology]. (4th ed.). München, Germany: J. A. Barth.

Wiener, N. (1948). *Cybernetics.* New York, NY: John Wiley & Sons.

Wigner, E. (1960). The unreasonable effectiveness of mathematics in the natural sciences. *Communications in Pure and Applied Mathematics, 13*(1), 1–14. (Richard Courant Lecture in Mathematical Sciences, delivered at New York University, May 11, 1959)

Wiles, A. (1995). Modular elliptic curves and Fermat's last theorem. *Annals of Mathematics, 141*(3), 443–551.

Whitrow, G. J. (1932). The importance of the history of mathematics in relation to the study of mathematical technique. *The Mathematical Gazette, 16*(220), 225–227.

Whitrow, G. J. (1933). Continuity and irrational number. *The Mathematical Gazette, 17*(224), 151–157.

Wold, H. (1938). *A study in the analysis of stationary time series.* Uppsala, Sweden: Almqvist and Wiksell.

Wold, H. (1949). Estimation of economic relationships. [Supplement: Report of the Washington Meeting (July, 1949)]. *Econometrica, 17,* 1–22.

Wold, H. (1954). *A study in the analysis of stationary time series* (2nd rev. ed.). Uppsala, Sweden: Almqvist and Wiksell.

Zeeman, E. C. (1962). The topology of the brain and visual perception. In M. K. Ford (Ed.), *Topology of 3-manifolds and related topics* (pp. 240–256). Englewood Cliffs, NJ: Prentice-Hall.

Zeeman, E. C. (1977). *Catastrophe theory: Selected papers, 1972–1977.* London, England: Addison-Wesley.

CHAPTER 3

THEORETICAL RECURSION IN RADICAL EMPIRICISM

The Universal Philosophical Acid

Eric P. Charles
Penn State Altoona

Recursion plays a crucial role in many theoretical approaches to psychology. This occurs most obviously in the "radical" approaches to psychology, which attempt to apply very simple and straightforward principles to understanding psychological phenomenon. We might count pragmatism, radical empiricism, and some forms of radical behaviorism among these approaches. The principles at the center of these systems tend to be uninterestingly mundane and intuitively acceptable when they are first put into use. However, when the principles are allowed to recurse—that is, when the principle is reapplied to the output of previous applications—very unintuitive results can arise very quickly. These unintuitive results are often the ones purported to have the greatest consequence for the field of psychology. Here we will try to explain how the recursion in William James's radical empiricism has consequences for psychology, and show how the same recursion is rephrased in E. B. Holt's brand of radical behaviorism.[1]

Cultural Psychology of Recursive Processes, pages 103–118

RADICAL EMPIRICISM

James's theory of radical empiricism is best expressed as an exclusive devotion to the principle that "Things are what you experience when you experience those things." This is about as simple a fundamental principle as I think any theory could offer, and it could be dismissed as an obvious truism, if so many approaches to philosophy and psychology did not argue for the antithesis. In particular, James's principle is antithetical to both idealism and dualism: The idealists hold that your experience is not of things. The dualists hold that the experienced things are never the things themselves. While there are many possible examples of the application of the radical empiricist principle in psychology, the truth is that it has not been developed to anything near its full potential. One extensive example is James's (1902) book on *The Variety of Religious Experiences* in which, among other things, the question "What is a religious experience?" is subjugated entirely to the question "What do people experience when they have religious experiences?" At the extreme, it has been claimed that radical empiricism provides a new foundation for not just psychology, but for all of science.

How can following such a seemingly simple principle be claimed to have such dramatic consequences? The answer is that the principle is applied recursively—faced with the answer to any question, the principle is used again and again and again—and in the process it becomes a sort of universal philosophical acid. The principle pulls every erected philosophical edifice of theory, no matter how convoluted and reified, back down to a first-order problem of understanding experience.

Simple Objects

The radical empiricist doctrine is most approachable when dealing with straightforward "things." Dewey (1905) gives an example of this in which many people, including a zoologist and a horse-dealer are asked to describe the same horse. Each person gives a different answer. Presumably, each is willing to argue that his answer is correct, i.e., that his is a better description of the "real" horse that that offered by the others, and that the other person was "wrong" in their judgment. Each would be willing to say, "My horse is real, yours is merely phenomenal." But Dewey admonishes us:

> If these accounts turn out different in some respects, as well as congruous in others, this is not reason for assuming the content of one to be exclusively 'real,' and that of others to be 'phenomenal'; for each account of what is experienced will manifest that it is the account *of* the horse-dealer, or *of* the zoologist, etc., and hence will give the conditions requisite for understanding the differences as well as the agreements.... In each case, the nub of the

> question is, *what sort of experience* is meant or indicated: a concrete and determinate experience, varying, when it varies, in specific real elements, and agreeing, when it agrees, in specific real elements, so that we have a contrast, not between *a* Reality, and various approximations to, or phenomenal representations of Reality, but between different reals of experience. And the reader is begged to bear in mind that from this standpoint, when 'an experience' or 'some sort of experience' is referred to, 'some thing' or 'some sort of thing' is always meant. (Dewey, 1905, pp. 393–394)

Let me try this with another example: What is a bowl of spaghetti? Well, what do you experience when you experience something as a bowl of spaghetti? Place a bowl in front of yourself and meditate for a moment. Though answers to the question could certainly go into more or less detail, depending on the person, and their dedication to the task, now let us say that we experience a bowl with long Italian-style noodles, about an eighth of an inch thick, with a savory tomato sauce. Well, there you have it—that is what a bowl of spaghetti *is.*

Presumably, these two examples are sufficient to get the gist of using the radical empiricist mantra. Inquiries into questions such as "What is a cup?", "What is a dog?", "What is a skylight?" and so on would all take the same form. In all cases we would look into our experience to determine what the thing is.

Here, perhaps, the dualist interjects, "But wait, you only experience the visual impression of things that might not at all be what they seem!" To which the radical empiricist responds, "Come now, come now, let's not be coy. Though you might later find yourself in error—perhaps what you saw was really an expertly prepared 'food sculpture'—that doesn't negate your initial experience. Haven't you ever learned about the psychologist's fallacy?" The dualist interjects again, "But wait, won't you now experience the very same object *as* a food sculpture! And might that experience not be further contradicted in the future? How can you claim such certainty in a world full of doubt?" To which the radical empiricist replies, "You're being silly again. Separate experiences are separate. Some new experience might replace the old one, true enough, but that doesn't nullify the fact that the original experience happened. All I am claiming is that each experience is what it is."

Needless to say, these arguments can get very convoluted very quickly, and my job here is not to defend radical empiricism against its possible critics. It should be sufficient here that each reader's inner seven-year-old agrees: Even if we later find out the bowl does not contain what we thought, we experienced it *as* a bowl of spaghetti when we experienced it as containing the noodles and sauce. Against the painful obviousness of that statement, it should again be noted that this line of reasoning is in complete and utter contrast to an analytic approach to answering the same questions.

What the radical empiricist will *not* do is to seek an authoritative source to tell us what to experience, or by which to judge our experiences right or wrong.[2] We did not look anything up in a dictionary, make reference to an absolute mind, discuss epistemically sound ways to judge the "truth" or "untruth," and so on. If a friend who is savvier about food walked by and said "Oh look, a nice plate of vermicelli," our future experience might or might not be altered, but either way our original experience was what it was, and should be treated as such. Spaghetti *is* what we experienced when we experienced the stuff in the bowl *as* spaghetti. Again, if it wasn't so controversial, we could call it a truism.

RECURSION—APPLYING RADICAL EMPIRICISM TO LESS "THINGY" THINGS

Though the history of philosophy makes the radical empiricist's proposition controversial even involving mundane physical objects, there is little obvious controversy there for the layman or the psychologist. The novelty of radical empiricism is not obvious until the principle is applied recursively, and there is little reason to apply it recursively until we start to apply the principle to things that are less concrete, less "thingy." To illustrate the variety of things we can apply this principle to, and thereby show the recursion in action, let us imagine a conversation with William James as the protagonist, with Jean Piaget, Bishop Berkeley, and Rene Descartes as the antagonists. For the sake of convenience, they will all talk in modern voices. I will attempt to give a more formal description of some of the steps afterwards:

Piaget: A child was shown a type of "chair" he had never seen before. It was constructed of several pieces of stone, and hence it had no legs. Despite the child's schema for chair including that the object must have four legs, the boy quickly accommodated his schema "chair" to include this new object.

James: But my dear Piaget, the empirical fact for the child is simply that the novel object is a chair. The rest of that describes *you.* It tells what is happening when *you* experience the child *as* accommodating their schemata. There is little evidence that the child experiences any part of that interaction as involving "mental schemata."

Piaget: Well of course they don't. My description of the child is more true of the child than is the child's own experience. My point of view *is* privileged.[3] Kant showed that percepts without concepts are blind, and I have spent many years studying how this epistemic riddle is solved through developmental processes. The child *must* have a preexisting mental category of "chair" into which the perceived object is placed. *I* know the child's category required the object to have four legs, and *I* know that this object does not have four legs, so if the child calls it a chair *de facto* he has adjusted his category to allow for non-legged objects!

James: But I beg you again to notice that what you have described is a truth about yourself. *You* experience the child as accommodating the new chair, because of the things present in *your* experience. But as your experience involves things not present in the child's experience, it would be absurd to argue that what you say is true of the child. This is the psychologist's fallacy. The child experiences the object as a chair, and the presence or absence of legs that you claim to have such importance, might have little importance to the child, even if he has never seen a legless chair before.

Piaget: But I have asked the child to describe chairs to me on many occasions, and the child has always told me that they have four legs. I am referring to the child's category, *not* my own.

James: Tut, tut, you change the subject. What the child experiences as "a chair," and what they experience as "a description of a chair" are two different things. There is no reason *a priori* to assume that the two types of experience are consistent in the way you imply. In fact, the existence of several interesting psychological phenomenon suggests that people are commonly inconsistent in just such a way. For that matter, why are you so confident that to experience something "as a chair" is to experience it "as a member of the category chair" or "as something that fits the definition chair"? Of course, sometimes the latter *are* experienced—for example, after reading a placard in an avant-garde art gallery one might experience the object before them specifically as "a member of the category chair," and similar experiences can be invoked during an academic argument, but they rarely occur in mundane interactions.

Berkeley: Sorry to interrupt Dr. James, but you are failing to make the crucial distinction that makes this conversation necessary. The child's conception of the object as a chair is a purely mental event, whereas the structure itself, the so-called chair, is physical.

James: I assure you dear Bishop, that such a distinction, crucial though it may be to some endeavors, is entirely *post hoc*, and not *a priori*. Only having experienced the chair can we have the additional experience that the chair *is* mental, or *is* physical. If we *then* want to know the difference between the mental and the physical, we must look *again* into our immediate experience for the answer.

Berkeley: But the child has not in his mind the chair itself; he has only the idea of a chair. Perhaps I was being to generous earlier; let me be more blunt: The notion that the child ever experiences the physical chair is an illusion that must be avoided at all costs. He only experiences the idea-chair. You cannot avoid this obvious conclusion with your hand waving, nor by bending the meanings of the words.

James: I am not bending the meaning of the terms. I am sticking up for their obvious and straightforward meaning. When I say that something is physical, I mean that I experience it as physical. When I say that something is mental, I mean that I experience it as mental. Given that, it is *not* mere hand waving to point out that questions regarding the nature of the mental and the nature of the physical will be found through a proper examination of those particular experiences.

Descartes: If I may, Dr. James, all this is fine and good, but surly you are aware of my simple proof that the only thing truly known is the mind itself.

James: Prof. Descartes, I respect the importance of your work in the history of philosophy, but I must say that it seems to have taken us far off the track. Your argument merely asserts several new varieties of experiences, and they *are* all worth examining, but we mustn't pretend that you did much more than that. It might be the case that everything *you* experience is experienced *as* mental, which seems doubtful, but I for one experience hardly any of my surroundings as mental on any given day. It is a strange arrogance that leads you to assume that your personal experience should be privileged over mine in the way you suggest. An even stranger arrogance leads you to privilege your armchair imaginings over much more common and pervasive everyday experiences.

Descartes: It is not imagining! It is *deduction* that leads me to these indisputable conclusions! I have found the truth, and you wish to deny it.

James: But here we are at the psychologist's fallacy again. You are asserting that something difficult to discover was really where you started. At any rate, your claim is experienced as true by some people and as fiction by others. It should thus be clear that "truth" and "fiction" also refer to two different ways of experiencing something. It is not likely that "deduction" restricts which things can be experienced as "true" and which as "fictitious," in the manner you suggest. I will grant, for I have seen students do it, that one can for a short time, or almost can, experience anything not deduced as fictitious. However, such exercises are extremely forced and short-lived. Further, when such an exercise is finished, nothing about it has suggested that it is "better" than the manners of being that preceded or followed it.

Berkeley: But Dr. James, it is still the case that all the child has is the idea of a chair. Descartes is correct, at least on this point: All the child knows for certain is that he has the idea of a chair!

James: No. The child has experienced *a chair*. If you deny that, what is there to discuss? The whole discussion is premised on the child experiencing a chair, and yet you do nothing but try to claim that the child experienced other than a chair. It is hogwash. *You* experience the chair as "a concept" or "an idea," but the child experiences a chair. Given time, you may also get the child to experience the chair as "a member of a category" or as "mental," but doing this will not tell you anything interesting about the original experience, nor prove any metaphysical point. At best it will tell us something about the developmental processes that alter experience.

All three antagonists: But how can you deny the distinction between the mental and the physical!

James: Because it is not true to experience, or at best it is true to only the smallest subset of experiences. I refuse to privilege some experiences over others in the manner you desire. The answer to all questions such as these, no matter how complex, is to be found through an examination of the immediate experiences to which they refer.

RADICAL EMPIRICISM AS PHILOSOPHICAL ACID

Hopefully by now it is clear that one consequence of taking the radical empiricist mantra seriously is that all questions, whether philosophical, scientific, or otherwise, become problems of understanding experience. In this way, the recursion acts as a theoretical acid, dissolving down all erected distinctions, such that they must first be understood in terms of immediate experience. This is not to say that we cannot build theoretical structures, only that those structures must stay true to experience, and if they are challenged at any time, the challenge must be answered in terms of experience.

In the mock conversation above, distinctions between "percepts" and "concepts," between "physical" and "mental," between "idea" and "object," between "truth" and "fiction" were all pulled down from their pedestals. When allowed to remain on their pedestals, such distinctions are held to be "true" about the world *before* the world was experienced; they are seen as categories that control and limit experience. However, when pulled down from their pedestals, each distinction is revealed to be a distinction between different ways of experiencing things; they are categories derived from experience. The way to properly investigate these distinctions is not by erecting shiny, logically pretty and foolproof edifices that keep them floating above us in the ether. All such terms are properly investigated not on a pedestal, but on the ground, within the muck of experience.

This might be better explained in an iterative example. I will start with a quick episode, presented as a standard, first-person narrative. Next I will analyze the story from both a traditional perspective and a radical empiricist perspective. The traditional perspective will take dualism for granted, as well as the rightness or wrongness of any judgment about the world. The radical empiricist perspective will simply examine the experiences themselves.

The Episode

It is dark, but I slowly become able to make out a form. It is a man. I call out, but get no reply. I approach, and squint. It is not a man; it is statue, a very good statue, maybe wax. I thought I saw a man, but I was wrong. It was only a man in my mind; the statue is real. Wait, now my eyes are opening again. It was all a dream. There was never anything there at all.

Traditional Dualistic Translation

This story is about a person doubly tricked. At first he thinks he is seeing a man, then that is replaced by thinking he is seeing a statue. In fact, there never was any such form anywhere. Everything that supposedly "happened" was merely in his head. Mid-dream, he was correct in asserting there was no man, but wrong in asserting there was a statue. He is correct only at the end, when he judges both objects to have never existed.

Radical Empiricist Translation

This story is about a person's transforming experiences. The form is experienced first as not having a clear shape, but then quickly comes to be distinguished as a man. Then the form is experienced as a statue. After the form is experienced as a statue, the original experience is re-experienced as wrong. After it is experienced as wrong, it is also experienced as having been mental. Then the person experiences all of those happenings as "mental" and the room he finds himself in as real. More specifically, the prior things are re-experienced as having been "dreamt" *and* as having been "mental," whereas the current surroundings are experienced as physical.

Elaboration of Radical Empiricist Translation

There are crucial differences between the radical empiricist translation and the traditional translation that are easy to miss. To highlight but a few: (1) In the traditional translation, the original experience of the man is declared to have been purely mental. In the radical empiricist translation, it is emphasized that no such distinction originally existed—there was nothing about the original experience to suggest that it was "wrong" or "mental." Those are aspects of new experiences, not the original experiences. (2) In the traditional translation, there is no *thing* being experienced. Part of what the dualist asserts by declaring something to be "mental" is that it is "not real." Even were we to somehow force the dualist to accept the dreamed form as "a thing," they would still insist that the experienced man was distinct from the experienced statue—that is, that there was one something originally and a different something later. The radical empiricist, on the other hand, accepts both the experienced form as a thing, and as the same thing despite the transformation. It is necessary to refer to the form as a stable thing, because a stable "sameness" was part of the dreamer's experience. (3) In the traditional translation, once everything is revealed to be a dream, this retroactively dictates our treatment of the original experiences as composed of "dream stuff" (be it ideas, misfiring neurons, illusion, or some other substance). In the radical empiricist translation, we stay true to the obvious fact that such is a *post hoc* judgment. Unless the original experience was somehow "dreamy" as, for example in the case of a lucid dream, it is a gross violation to treat the original experience as somehow having been of "dream stuff." The *last* experience is of the previous experiences *as* dreams—that is, the last experience only.

To focus on the final point: If we want to understand the difference between "dream" and "real," we need to look at the difference between the original experience (of the statue as "real statue") and the last experience (of the statue as "dream statue"). Whatever is different between those two concrete experiences is the meaning of "dream." It does no good to simply declare that the *first* experience was of "dream statue"; in fact, to do so is to completely undercut our ability to investigate the phenomenon of interest.[4]

The radical empiricist stays true to experiences in ways that the traditional approach does not. The original experience was *not* of a "real statue" nor of a "dream statue," but merely of "statue." In this sense, the original experience was neutral with respect to that distinction (see Dewey, 1917). As we found in our multi-philosopher discussion above, we again find that all categories are *post hoc,* in that they are part of a later re-experience. However—and here is the recursion—those later re-experiences are also themselves experiences. Thus, the re-experience must be subjected to the same analysis as the original experience. The categories revealed in our re-experience are themselves first-order members of the particular experience in which they are found. No amount of compounding experiences can escape this. It is not that we are getting nowhere with our thinking, re-thinking, or meta-thinking; it is only that wherever we get, we are still within the realm of pure experience.

EXPLICIT RECURSION

Alas, James did not fully develop his radical empiricism before his death, so we have few examples of it being applied in his hand. The instances that exist, however, are rich. Let us take, for instance, James's (1904) discussion of the overlap between his position and the ideas being developed by Dewey's "Chicago School" of philosophy. Here James distinguishes between theory and fact based on their function in experience, and rejects the notion that they are distinguished by comparison to any extra-experiential structure:

> What is a fact? A fact and a theory have not different natures, as is usually supposed, the one being objective, the other subjective. They are both made of the same material, experience-material namely, and their difference relates to their way of functioning solely. What is fact for one epoch, or for one inquirer, is theory for another epoch or another inquirer. It is 'fact' when it functions steadily; it is 'theory' when we hesitate. 'Truth' is thus in process of formation like all other things. It consists not in conformity or correspondence with an externally fixed archetype or model. Such a thing would be irrelevant even if we knew it to exist. Truth consists in a character inclosed within the 'situation.' Whenever a situation has the maximum stability, and seems most satisfactory to its own subject-factor [i.e., person], it is true for him.... Experience is continually enlarging, and the object-factors [i.e., things] of our situations are always getting problematic, making old truths unsatisfactory, and obliging new ones to be found.... [T]he truth creates itself *pari passu* [in lockstep with our ongoing experience], and there is no eternally standing system of extra-subjective verity to which our judgments, ideally and in advance of the facts, are obliged to conform. (James, 1904, p. 4)

In that same year, James would put forth one of his first explicit descriptions of his radical empiricism (James, 1904/1996a). James begins with a description of how he would deal with mundane, immediately present, physical objects, and then moves on to deal with concepts and remote objects:

> So far, all seems plain sailing, but my thesis will probably grow less plausible to the reader when I pass from percepts to concepts, or from the case of things presented to that of things remote. I believe, nevertheless, that here also the same law holds good. If we take conceptual manifolds, or memories, or fancies, they also are in their first intention mere bits of pure experience, and, as such, are single *thats* which act in one context as objects, and in another context figure as mental states. By taking them in their first intention, I mean ignoring their relation to possible perceptual experiences. (p. 7)

Shortly after this, James starts speaking of the different *thats* being in relation to each other. "Relations" are a new element of his system, but they too are subject to the same analysis. Though it may have initially seemed as if James had identified a higher-order structure that moved beyond an individual experience, the notion of extra-experiential hierarchy is quickly dissolved. That is, James's analysis of experience shows the existence to contain relations, and faced with this realization, James reapplies the very same method of analysis, asking what a relation is when viewed as a first-order element of experience. The recursion is shown in a footnote: "Here as elsewhere the relations are of course *experienced* relations, members of the same originally chaotic manifold of non-perceptual experience of which the related terms themselves are parts" (p. 7). This is particularly important, as James will later claim that "knowing" is a kind of relation, one of many ways of experiencing things. Thus, "knowing" as well as "the knower" and "the known" (the two things related in the act of knowing) are all identified as additional experienced elements of a situation. This analysis of the 'cognitive relation' is continued in James's "The World of Pure Experience," which came out later in the same year (1904/1996b).

These papers show James's recursive use of his core principle: "Things are what you experience when you experience those things." Asked to answer any question, he reapplies the principle, until he has found first-order, concrete experiences to which the question refers. Each specialized term, each higher-order concept, each seemingly "objective" utterance, is interrogated, and re-interrogated if necessary, until it place is found in the immediate experience of some person in some situation.

RECURSION IN SOME FORMS OF RADICAL BEHAVIORISM

There are many forms of radical behaviorism. Some of those forms are intellectual descendents of radical empiricism. This is most obvious in the case of E. B. Holt's new realism. Holt was a disciple, friend, and protégé of William James, who saw himself as continuing the tradition of radical empiricism (Charles, in press). After James's death in 1910, Holt's goal was to develop the tradition in the way James would have if alive (e.g., Holt, 1914, 1915, 1931). How, then, does Holt end up as a radical behaviorist? Dewey has already, to an extent, shown us the way. At the end of his discussion of the horse, Dewey tells us that "the reader is begged to bear in mind that from this standpoint, when 'an experience' or 'some sort of experience' is referred to, 'some thing' or 'some sort of thing' is always meant" (1905, p. 394). Further, James has told us that the cognitive relation is understandable as a concrete way in which knower and known relate. Combining Dewey's and James's points, we can readily transform a radical empiricist statement into a radical behaviorist statement: The radical empiricist asserts

> Another person's mind is whatever you experience when you experience that person's mind.

Applying Dewey, we substitute all references to "experiences" for references to "some thing" and find that

> Another person's mind is the *things* you see when you experience their mind.

Applying James, we note that minds are nothing other than concrete relations of some sort, and find that

> Another person's mind is *a thing you* see when you see a *particular type of concrete relation* between the knower and the known.

We can generalize this a touch to

> Another person's mind is nothing more than a particular type of observable relation between observable things (i.e., a knower and the known).

If we then ask ourselves, *what observable things are there to see regarding the knower and the known?* The answer is clear: *We see behaviors relative to their circumstances.* And just like that, we are suddenly advancing some form of behaviorism:

> When we see another person's mind, we are seeing some pattern of behaviors relative to circumstances. And if that is what we see when we see someone else's mind, then that is what the mind is. Period.

So, for example, if we were trying to understand Piaget's notion of "egocentrism," which bears striking similarity to the modern notion of a person lacking "theory of mind," how might we do so? The traditional approach would have us start by asking what it means to be "egocentric" or what it means to not have a "theory of mind." We would define *a priori* the cognitive ability that a person might or might not have, and then we would go about devising tasks to see if the people in question have that ability. Our *a priori* definition and our predesigned laboratory tasks will then be taken as the *sine qua non* authority for what, exactly, we are talking about. The radical empiricist, in contrast, would ask us to compare what we experience when we experience someone *as* egocentric with what we experience when we experience someone *as not* egocentric. That is, we would look to the *immediate experience* as the ultimate authority for what we are talking about. Holt would quickly point out that what we experience are certain behaviors under certain circumstances. This is the transition from radical empiricism to philosophical behaviorism. If we can carefully determine the types of behaviors-relative-to-circumstances that constitute our experiences of *someone in a particular circumstance* as egocentric, then we can come to a consensus about what egocentrism *is*, and how to determine whether someone else has it. The answers are still found by looking at the things experienced, but with an added caveat: When we examine our experiences of other people's psychological states, the things experienced are behaviors.

This leads Holt to his major metaphors of consciousness, the searchlight metaphor and the cross-section metaphor (Holt, 1914). Just as James (1904/1996b) points out that something can be "mind" in one context and "matter" in another context, Holt tells us that the world *sliced* one way reveals certain properties of an object while the world *sliced* another way reveals different properties.

This form of radical behaviorism then relies on recursion of its fundamental methodological principle in the same way that radical empiricism does. Given an assertion regarding a psychological state, Holt asks what was seen. If the answer to that question reveals a new host of psychological terms, Holt interrogates each of them by asking again, what was seen. This process continues until you are left with individual people, behaving, in some circumstances, or patterns of behavior across an array of circumstances. That is, whenever faced with the need to analyze a new phenomenon, we ask what aspects of the world we are *responding to* when we respond to that thing.

A FINAL EXAMPLE—UNDERSTANDING SCIENTIFIC DISCOVERY

Let us say that we are interested in understanding the process of science, and particularly the phenomenon of "scientific discovery." There are, of course, many traditional ways of trying to get at this notion, to determine definitively what is, or is not, "a discovery." We might go with the "history of science" discipline and construct a definition via induction, collecting historic data about widely agreed upon "discoveries" and then trying to determine the key commonalities across those historic situations (for example, we note that evaluations of originality are culturally limited, as many well-agreed upon European "discoveries" involve ideas the Chinese and Inca had both "discovered" several hundred years earlier). Alternatively, we might go with the "philosophy of science" discipline and construct a definition of "discovery" via deduction—that is, we could begin with a set of postulates and construct a rigorous proof that would determine whether or not a given incident constitutes a discovery (for example, we could simply declare that something can only be "discovered" once). The approach required by radical empiricism is completely different. The radical empiricist would tell us: If you want to know what a "scientific discovery" is, you need to look into the experiences of someone who is experiencing a given bit of data *as* a scientific discovery.

Or, as Holt would put it: When you slice the world one way you see the features that make X a scientific discovery, when you slice it another way you see the features that make X a good candidate for the shredder. There is no logical sense in which what one person sees can "trump" what another person sees in the manner traditional theories suggest. Both the person who sees X as "a scientific discovery" and the person who sees X as "not a scientific discovery" are responding to real aspects of the world. If we are trying to understand what "scientific discovery" *is*, then our task is to determine what aspects of the world are being responded to.

CONCLUSIONS

I certainly do not wish to give the impression that what I have said above is all there is to radical empiricism or philosophical behaviorism. William James's thinking is well reputed for its amazing breadth, and those few who have followed in his foot steps deserve more attention than they have received. The purpose here was to demonstrate an important use of recursivity in psychological theory; as such, that particular aspect of radical empiricism was focused upon.

When discussing James, or others who promote unusual theories, it is often said that they were intellectually "brave." What is this braveness that is spoken of? My *experience* tells me that it is a willingness to see ideas through to their logical conclusions without taking easy outs, and especially without seeking escape into superfluous or contradictory ideas. One way of detecting this braveness is to see if the people are willing to recurse their logic, to apply their methods in critical examination of their own results or cherished beliefs. Other people's *experiences* may be different; some others seem to experience intellectual braveness as primarily a willingness to say things that are unpopular, or even just a willingness to say a particular kind of thing. Thus, some people seem to think a person is brave if they are willing to declare that "All findings are culturally relative." I would not experience that same person as intellectually brave unless I saw them recurse their logic and come to terms with the consequence—their own declaration must also be viewed as culturally relative. Similarly, we may expect of the radical behaviorist an admittance that their own behavior in writing an argument is analyzable in the same manner as any other behavior they may wish to study. Thus, at least for some, a willingness to pursue intellectual recursion is seen as a high virtue.

NOTES

1. I should admit from the start that this is not an exceptional piece of scholarship, in the classic sense. Though citations are included, this work is more like a scholarly essay than a research paper. The goal is to get a complex idea across to the reader, rather than to establish the historic basis of this line of thinking, or acknowledge all who have published similar ideas. I thank the editor for allowing me the indulgence to focus my paper in this manner.
2. Of course, we might do that for other purposes, but not for this purpose.
3. While I doubt Piaget would have made this claim so explicitly, it must be what he would ultimately claim. Surely he thinks his description of the child is more accurate than the child's self-report would be, or there is no need for him to create these theories. As I am emphasizing the ideas, rather than the actor's self-presentation, hopefully fans of Piaget will forgive the brusque tone.
4. Though I will not elaborate on it here, one can also see at this point the necessity of James's pluralism, as there is no reason to believe that "dream," when investigated in this manner, will have identical meanings between people or between instances within the same person. We would, however, expect some degree of consistency across large swaths of people, and across many experiences within the same person.

REFERENCES

Charles, E. P. (2014). William James's personal and professional influence on Edwin Bissell Holt: A student, colleague, disciple, and friend. *William James Studies, 10.* (http://journal.wjsociety.org/?name=535)

Dewey, J. (1905). The postulate of immediate empiricism. *The Journal of Philosophy, Psychology, and Scientific Methods, 2,* 393–399.

Dewey, J. (1917). The concept of the neutral in recent epistemology. *The Journal of Philosophy, Psychology, and Scientific Methods, 14,* 161–163.

Holt, E. B. (1914). *The concept of consciousness.* London, UK: George Allen & Co.
Holt, E. B. (1915). *The Freudian wish and its place in ethics.* New York, NY: Henry Holt.
Holt, E. B. (1931). *Animal drive and the learning process: An essay toward radical empiricism* (Vol. 1). New York, NY: Henry Holt.
James, W. (1902). *The varieties of religious experience.* New York, NY: Longmans.
James, W. (1904). The Chicago School. *Psychological Bulletin, 1,* 1–5.
James, W. (1996a). Does consciousness exist? In E. I. Taylor & R. H. Wozniak (Eds.), *Pure experience: The Response to William James.* (pp. 1–17). Bristol, England: Thoemmes Press. (Original work published in 1904)
James, W. (1996b). A world of pure experience. In E. I. Taylor & R. H. Wozniak (Eds.), *Pure experience: The response to William James* (pp. 18–40). Bristol, UK: Thoemmes Press. (Original work published in 1904)

CHAPTER 4

"IMAGINED RECURSIVITY" AND STIGMA MANAGEMENT AMONG AMERICAN ATHEISTS*

Brooke Long
Fritz Yarrison
Kent State University

Nicholas J. Rowland
The Pennsylvania State University

ABSTRACT

Recursivity is a concept explicitly used in mathematics and linguistics; however, the term has yet to find a stable home in the social sciences. The purpose of this chapter is to contribute a social psychological variant of recursion, mainly, "imagined recursivity," by returning to the work of Cooley and the "looking-glass self." We examine the role of imagined recursivity in stigma management, in particular, regarding self-disclosure. We draw from 35 in-depth interviews with a growing group in the United States called "Nones" (i.e., individuals who do not self-identify with any religion). Respondents claim that if plainly asked in conversation "are you religious?" they would

Cultural Psychology of Recursive Processes, pages 119–142

openly reveal their ("discreditable") nonreligious identity, which we will call the "displaced burden of disclosure." Additionally, we believe that this technique has the practical utility of "hiding without hiding" their nonreligious identity. This protects the identity's owner from feeling the need to actively conceal their identity under the promise that if they were hypothetically asked they would hypothetically tell. This strategy is, above all, an interactional gambit; in exchange for the freedom of not feeling silenced, individuals risk the chance to be revealed.

INTRODUCTION

This book chapter makes two contributions. The first is conceptual. We contribute "imagined recursion," which is a social psychological conceptualization of recursion that has implications for the relationship between recursion and human behavior. The chapter begins by reviewing definitions of recursivity currently used in mathematics, linguistics, and the humanities, and then searches basic literature in social psychology for recursion. We emphasize early social psychological literature rather than contemporary research in attempts to capture a core notion of recursion as it applies to social psychology. As it happens, recursion is a term that is not formally defined or explicitly used in social psychology. However, as we shall see, it appears to be at the center of one of social psychology's most basic concepts, Cooley's (1902) "looking-glass self," from which we develop "imagined recursivity."

The second is empirical. This contribution involves an analysis of in-depth interviews. We examine one social stigma management strategy employed by "Nones"—individuals who identify with no religion—to negotiate their common nonreligious identity during interactions. We find that the most common strategy adopted by respondents predates and partially obviates certain face-to-face interactions, which they reveal is largely contingent upon imaginary interactions that rarely, if ever, occur. In what we will call the "displaced burden of disclosure," respondents state that they would be willing to disclose their ("discreditable") identity as a nonreligious person, but only if asked directly during a face-to-face interaction. This technique of passively avoiding identity disclosure during interactions, based largely on imagined scenarios of how reality will unfold, effectively acts as a passing mechanism with the practical consequence of delaying if not obviating disclosure. Additionally, this technique, which we believe is a new contribution to the stigma neutralization literature, has the pragmatic utility of "hiding without hiding" their nonreligious identity, which protects the identity's owner from feeling the need to actively conceal their identity under the promise that when asked they would tell. Our respondents do not necessarily see themselves as silenced; it is just that they are infrequently asked about being nonreligious. This strategy is, above all, an interactional gambit; in exchange for the freedom of not feeling silenced, individuals risk the chance to be revealed.

RECURSIVITY IN LITERATURE

Recursivity is a term mainly used in mathematics, linguistics, and the humanities.

Our use of recursion in this chapter, however, is largely based on the definition of recursion in the humanities where recursion "refers to the act of a mind or self 'looping back,' 'turning around' or reflecting on itself, and in this way actually creating itself as a conscious self—the highest expression of human awareness" (Clifford, Friesen, & Jardine, 2001, p. 200).

This definition melds with a basic assumption about humans and their ability to think recursively—"that they can and often do have thoughts about the thoughts or mental states of others" (Levy, 2009, p. 372). The definition of recursion from the humanities finds good fodder in studies of social interaction—in particular, the core notion in symbolic interactionism, mainly, the "looking-glass self" (Cooley, 1902) along with stigma management techniques (Falk, 2001; Goffman, 1986; Heatherton, Kleck, Hebl, & Hull, 2003; Kroeger, 2004).

In general, recursive processes are characterized by continuous looping; however, loops are never identical even if they appear so and are, therefore, in a state of constant flux. Recursion is also not merely imitation as such would imply exact sameness of every repetition. Rather, given that slight differences within each loop exist, in principal, recursion is better understood as a process of emulation (Hurley, 2008). Recursive processes are not repeatedly mirrored; instead, loops appear to be modeled after one another but not necessarily building on one another toward some predictable or "better" end (Harries-Jones, 1995).

Human perception is central to understandings of recursivity. In the process of actively living our lives, we find ourselves unable to critically analyze seemingly insignificant differences amid the infinite nuances of any human endeavor. Therefore, we impose categories on the minutiae that constitute everyday life. Phenomena appear to be either similar enough to or different enough from those that precede it to be perceived and therefore categorized as such. Occasionally, minor changes characteristic of routine behavior, which emerge slowly over time, result in a form of incremental change that is not immediately observable by the actor conducting the behavior. The net result of many minor but incremental changes often goes unnoticed, but occasionally those changes have shocking if not disastrous consequences (cf. Vaughan, 1996). Even appropriate adaptive behavior may result in pathological looping such as when individuals use alcohol as a basic coping strategy to deal with moderate levels of stress. However, over time, this reasonably harmless adaptive behavior has the potential to create a tolerance to alcohol that, at a certain threshold, results in pathological alcohol use to recreate the same relief (Harries-Jones, 1995).

THE SOCIOLOGICAL SIGNIFICANCE OF (IMAGINED) RECURSIVITY

Conceptually, recursion is at the heart of social science, but formally, it does not appear to be a central concept in its disciplines. To support this statement, we reviewed 40 introductory textbooks in social psychology, sociology, and psychology for the explicit use of the term "recursion" (see "List of Introductory Texts" in the reference section). Introductory textbooks were purposely selected as a set of texts in our attempt to examine issues so salient to these related fields that they would enter into the basic lexicon of introductory texts. Book indexes were reviewed for the terms "recursion," "recur," and "recursive" as well as "iteration," "iterate," and "iterative." After examining the textbooks only one explicitly used the term "recursion." In *Cognitive Psychology* (2009), recursion refers to the linguistic sense of the term (i.e., that the generative capacity of syntax implies that words can be endlessly combined into new combinations, which constitutes a form of linguistic recursion).

However, recursion is relevant in the social sciences. In fact, we contend that recursion is so central to the field that it remains unintentionally under-explicated in core secondary texts such as textbooks. We are emboldened to take this stance by the inclusion of the "looking-glass self" in each of these many disciplinary textbooks and the obvious importance of recursivity for Cooley's core concept and his vision of individuals and society.

There are many concepts supporting the foundation of the social sciences. Specifically relevant to social psychological literature and research, is Mead's *Mind, Self, and Society* (1934), which describes a reflexive process between self, society, and behavior. This process is one such that society impacts self which in turn impacts social interaction creating a continuous relationship between an individual and the society in which they live. Mead, however, is not the only early theorist to describe a recursive relationship central to social psychological research and theory. Cooley, another of the most influential founding theorists, contributes to this discussion with his concept of the looking glass self. Based on a recursive process, Cooley's (1902) "looking-glass self" has "three principal elements: the imagination of our appearance, the imagination of his judgment of that appearance, and some sort of self-feeling, such as pride or mortification" (p. 184). This concept refers to social phenomena wherein individuals utilize the human capacity for recursive thinking. They imagine their appearance or the appearance of their behavior as displayed to others in a social situation. In particular, they imagine the sorts of judgments that individuals might state openly or think privately in response to their displayed appearance or behavior. All of this is followed by the individual estimating a personal/emotional feeling associated with the imagined appearance or behavior and the imagined response of others.

Since its construction in 1902, the looking-glass self has been researched, reworded, recast, reinterpreted, and reapplied throughout the social sciences. For the most part the idea has stayed within the realm of sociology (Yeung & Martin, 2003) and social psychology (Gecas & Schwalbe, 1983). Some research has attempted to combine Cooley's concept with others such as Scheff's discussion of the looking-glass self and its use in Goffman's symbolic interactionism (Scheff, 2005). As a concept, however, the looking-glass self has not progressed much further than its original articulation by Cooley. Most research that invokes the concept simply restates Cooley's idea and applies it to contemporary examples such as Schaffer's use of the looking glass self in the discussion of the evolution of religion (Shaffer, 2008). Hence, Cooley's concept appears to be shared by the social sciences yet unchanged.

Recursive thinking and social imagination also characterize Cooley's "self-idea." He states:

> [I]n a very large and interesting class of cases the social reference takes the form of a somewhat definite imagination of how one's self—that is any idea he appropriates—appears in a particular mind . . . so in imagination we perceive in another's mind some thought of our appearance, manners, aims, deeds, character, friends, and so on, and are variously affected by it. (1902, p. 183)

More broadly, imagination plays a prominent role in Cooley's (1902) thinking in *Human Nature and the Social Order.* Of "imagination as a whole," Cooley (1902) writes, imagination "is human thought considered in the largest way as having growth and organization extending through the ages"; imagination "is the locus of society in the widest possible sense" (p. 134). Further, Cooley (1902) observes that imagination is perpetually recursive in that "we always imagine, and in imagining share, the judgments of the other mind" (p. 184). Thus, if recursion is at the core of Cooley's looking-glass self and this concept is widely shared in the social sciences, then a notion of "imagined recursivity" characterizes what might be called a social psychological variant of recursivity—even if this vision of recursion is deeply taken for granted in the social sciences. A more contemporary example of this taken for granted nature can be found in recent work within symbolic interactionism (Stryker, 1980; Serpe & Stryker, 2011). Stryker (1980) extrapolates on early social psychological theory including Cooley and Mead to explore the relationship between self and society. This work, like Mead's (1934), discusses the reflexive nature of the self. By this, Stryker (1980), and later Serpe and Stryker (2011), refers to a constant process within which society impacts and individual's self, while at the same time, each individual self is impacting society. Thus, the subtle yet unnamed use of recursion by current symbolic interactionists, coupled with the obvious use of an "imagined recursivity" in the early social psychological work of Cooley, suggests that the notion of recursion is deeply ingrained in the social sciences.

This vision of recursivity in the social sciences is different from those definitions currently in use in other disciplines. In mathematics, recursivity refers to iterative processes, which build on one another not necessarily in a linear fashion but which can be worked-backward to a base case (i.e., the case of origin). Thus, mathematics uses an iterative recursive concept. In linguistics, recursivity refers to the capacity of human communication to embed phrases within phrases, wherefrom the nearly infinite complexity of interaction is born. Thus, linguistics also uses an iterative recursive concept, but one based on the embeddedness of language.

In contrast, recursivity, as it is used in the humanities, refers specifically to self-reflection or self-reflexivity wherein one's self can "loop-back" and thereby reflect on itself. Hence, the humanities also use an iterative model of recursion, but one explicitly related to the self reflecting on the self. Cooley also uses an iterative model of recursion related to the self reflecting on the self; however, in true social psychological form, this recursive process is one where the self reflects on the self as others see it. For Cooley, one way to understand the self is to recognize that individuals see themselves as others see them and, in the process of everyday life, tailor many of their actions to their interpretations of others' interpretations of them. In effect, individuals reflect on the self with emphasis on how others see it. Moreover, Cooley's implicit use of recursivity explicitly incorporates the imagined nature of the self as it would operate in social interactions that have not yet unfolded. Furthermore, Cooley adds specificity to one definition of recursion that explicitly evokes imagination. Now, imagined or not, such recursive interpretations take on a reality and gravity when used to guide human behavior—and this should be empirically studied. Thus, the contribution to recursivity studies is to recognize the deeply imagined nature of a social psychological definition of recursion. However, we want to be clear that we are discussing the sociological significance of imagination rather than the significance of the sociological imagination.

RECURSIVITY, RELIGION, AND IDENTITY

This chapter also rests on the premise that identity can be conceptualized as an internal resolution (personal identity), which is assigned an external label in certain social situations (social identity) (Cooley 1902). One's personal and social sense of self is taken here to be a perpetually emergent construction rather than one that is fixed or immutable (Dillon, 1999). As such, this implies that identities are not necessarily stable and instead exist (i.e., take on personal and/or social meaning) during interactions in different social settings where identities are simultaneously responded to, tested, and revised (Goffman, 1959; Hewitt, 2000). In addition to situation-specific interactions, broader sociohistorical shifts or events are expected to influence identity construction and management of the self (Peek, 2005).

For many, religion is a primary element of the self. Religious affiliation is expected to increase when organized religions also provide members with additional benefits indirectly related to religion or belief such as access to social capital and related economic opportunities (Hurh & Kim, 1990). For many, revealing one's religious identity among members of similar faith is a source of social solidarity binding individuals together, and sociologists have overwhelmingly conceptualized religion this way (Warner, 1993). However, for others who would disclose a nonreligious identity or one's personal dedication to godlessness, revealing one's identity can be socially isolating and potentially stigmatizing.

Being nonreligious is complex. On the one hand, being nonreligious is a personal identity; it is an internal state, which is not necessarily public and instead can be thought of as "meanings attributed to the self by the actor... asserted during the course of interaction" (Snow & Anderson, 1987, p. 1347). On the other hand, being nonreligious is also a social identity. In distinguishing between religious and nonreligious individuals, society creates groups of people that belong together and others who are targeted for exclusion; that is, they create a symbolic boundary between the groups (Epstein, 1988; Lamont & Fournier, 1992). Among sociologists, symbolic boundaries along religious lines have typically been interpreted as sources of inclusion rather than differentiation (Warner, 1993). In contrast, Edgell, Gerteis, and Hartmann (2006) ask if "Americans make invidious distinctions between believers and nonbelievers" (p. 211). Based on national survey data, they report that Americans "name atheists as those least likely to share their vision of American society" as compared to other groups such as Muslims, homosexuals, and immigrants (Edgell et al., 2006, p. 212). The authors suggest that as Americans became more tolerant of other religious groups over the last fifty years, increased religious solidarity is thought to have heightened the salience of symbolic boundaries distinguishing religious from nonreligious Americans. In a study of active atheists (i.e., atheists who belong to atheist groups or clubs), Hunsberger and Altemeyer (2006) found that more than 50% of respondents "reported that being a nonbeliever had produced difficulty with relatives and friends" (p. 55). Hunsberger and Altemeyer (2006) also collected questionnaires from 253 atheists in the San Francisco Bay area,

53% of whom claimed interpersonal difficulties regarding their internal resolution to be atheistic. They also obtained responses from 35 atheists in Alabama and Idaho, with over half of these individuals claiming similar social difficulties with family and friends. When asked why anyone would adopt such a discrediting identity, individuals responded that the alternative (i.e., a religious identity) was logically untenable.

STIGMA AND STIGMA MANAGEMENT

Based on quantitative and qualitative research, nonreligious identity appears to be a discrediting mark that reduces an individual "from a whole and usual person to a tainted, discounted one" (Goffman, 1963, p. 3). Stigmatization and stigma have been defined in many ways. Cocker, Major, and Steele (1998) emphasize devaluation as an outcome of stigmatization where "stigmatized individuals possess (or are believed to possess) some attribute, or characteristic, that conveys a social identity that is devalued in a particular social context" (p. 505). In their review article, Link and Phelan (2001) suggest that social scientists who study stigma tend to do so by prioritizing social theories above the lived experiences of the individuals they purportedly report on. Schneider (1988), for example, finds that researchers emphasize "their scientific theories and research techniques rather than to the words and perceptions of the people they study" (p. 64). Link and Phelan (2001, p. 367) "apply the term stigma when elements of labeling, stereotyping, separation, status loss, and discrimination co-occur in a power situation that allows the components of stigma to unfold." Stigmatized identities are "discredited" when they are visually confirmable, or "discreditable" when they must be revealed (Goffman, 1963, p. 4). The nonreligious identity is categorized as discreditable and the managed release of discreditable information during social interaction is known as "stigma management."

NEUTRALIZATION AND ACCOUNTS AS CULTURAL RESOURCE MANAGEMENT TOOLKIT

In social movements theory, the notion of a cultural tool box is a dominant theme. Broadly, Swidler (1986) suggests that culture, which is conceptualized as multitier, is composed of stories and symbols that, when combined, constitute a collective cultural "toolkit." Different cultures at different points in history tend to favor some elements or sets of elements within these shared toolkits. Moreover, individuals do not necessarily have equal access to use or employ these tools in certain situations or during certain time periods.

Understanding how and when individuals employ cultural resources to neutralize stigma is a fruitful avenue for research. Techniques that legitimately neutralize stigma include (1) denial of responsibility, when one is the victim of circumstances one cannot control and is, therefore, not responsible for; (2) denial of injury, when one insists no harm was the outcome of his actions; (3) denial of the victim, or the notion that whoever the victim(s), they deserved whatever they got; (4) condemnation of the condemners, shifting personal blame by emphasizing that those condemning the actions or identity are no better and, therefore, in no position to judge; (5) appeal to higher loyalties, when individuals displace the stigma assigned to their identity or action by appealing to the "greater good" such as loyalty to family and friends (Sykes & Matza, 1957; Thompson & Harred, 1992). All of these strategies are documented during interactions and as such fail to harness pre-interaction techniques designed to avoid such interactions in the first place. This point we develop in analysis.

A CLOSER LOOK AT SPOILED IDENTITY MANAGEMENT: THE CASE OF THE "NONES"

In order to explore and discuss recursion within the social sciences, face-to-face interviews with nonreligious individuals are interpreted and employed. These interviews provide a clear example of a recursive process (similar to the looking-glass self) in which individuals imagine how various scenarios will unfold, make a behavioral decision based on this imagined experience, and then actually perform the behavior, which in turn shapes the current situation.

Data, Data Collection, Method, and Analysis

This chapter uses an inductive approach to data analysis. Interviews were transcribed verbatim and coded for analytic themes. Early themes were used to direct the shape of later interview questions, the literature review, and follow-up interviews. All of this implies that the techniques for the development of grounded theory were followed (Glaser & Strauss, 1967).

Data came from face-to-face interviews with 26 men and 9 women. Participants were white and were, on average, 28 years old. Interviews lasted approximately 70 minutes; the shortest was 40 minutes while the longest was 160 minutes.

Participants were initially recruited through a freethinking group on a Northeastern college campus. This freethinking group is a student organization established and run by students enrolled at the university. Extensive literature exists about going off to college and losing faith (Hunsberger, 1978). More than other stages in the life course, individuals formulate nonreligious beliefs during adolescence and develop the underpinnings of later identity transformation (Hunsberger, 1978). However, through the use of fliers, and as the word spread to freethinking individuals not affiliated with the student organization, non-college-aged students contributed to the study.

Background on Nones

The U.S. is increasingly nonreligious. By 2008, one in five American adults reported no religious identity, according to the American Religious Identification Survey (Kosmin & Keysar, 2009). In 2011, there were between 3.6 and 5.2 million individuals who classify as theological atheists in the United States (Cragun, Kosmin, Keysar, Hammer, & Nielsen, 2012). A group that has doubled in number since the early 1990s, "Nones" are individuals that when asked the open-ended question "What is your religion, if any?" answered "None, No religion, Humanistic, Ethical Culture, Agnostic, Atheist, Secular" or "Don't Know" (Kosmin & Keysar, 2009, pp. 2, 23). The group is young, with 70% of Nones under 50 years of age (Kosmin & Keysar, 2009). From 1990 to 2008, the group grew by 138% and recorded the highest estimated net population growth as compared to any other religious identity (Kosmin & Keysar, 2009). This implies to Kosmin and Keysar that "the historic reluctance of Americans to self-identify… [with no religious affiliation in a telephone survey] seems to have diminished" (2009, p. 7).

Sources of this transition are multifaceted. The best predictor of one's nonreligious identity is being raised in a nonreligious family (Spilka, Hood, Hunsberger, & Gorsuch, 2003). Among individuals raised in religious families, Hunsberger and Altemeyer (1997) find that one's dedication to truth and personal integrity, which was originally learned through religious devotion, becomes the catalyst pushing one away from religious and toward nonreligious ways of life. When doubts loom too large to be reconciled within a religious sense of self, in an effort to maintain the integrity their previously religious background required of them, individuals search for new identities with more ontologically acceptable answers. In broader analyses of poll data, Hout and Fischer (2002) also show compelling evidence that the Nones emerged as a collective reaction to the fundamentalist ideals and political behaviors of the "religious right." Intuitively, individuals with weak ties to organized religion grew increasingly distant from the church in response to rising fundamentalism among the faithful and gradually composed the loose-knit group that social scientists labeled Nones. Similarly, Altemeyer (2004, p. 83) finds that parents are more likely to have "de-emphasized the family religion" if "hypocrisy among their fellow church members" is observed. Qualitative data in follow-up research show that individuals who became less religious because of the hypocrisy of fellow churchgoers reported, in general, that the behavior and beliefs of "everybody who went to church" or "so-called Christians" were conspicuously inconsistent (i.e., "saying one thing, doing another") (Altemeyer, 2004, p. 85).

RESULTS

This analysis section describes how discussing nonreligion is non-normative, and with this knowledge in combination with imagined interactions, religiously unaffiliated individuals insist on making the gambit that if asked in the course of casual conversation that they would reveal their otherwise obscured nonreligious identity.

Not Part of Commonplace Conversation

The norm is that religion, and nonreligion in particular, is not talked about during commonplace conversation. In general, our respondents suggested that it is normative, if not polite, to rarely, if ever, discuss religious and nonreligious ideas in public settings. George, a 20-year-old agnostic, suggests that civil conversation does not include religion. He states: "It is easier just to avoid the topic because... you can have a civil conversation without putting religion into it." Sally, a 19-year-old atheist, never remembers a time when nonreligion was mentioned in the course of casual conversation. In her words: "There are just never situations that I am in that it [i.e., nonreligion] comes up in." In reference to putting his best foot forward, Charles, a 19-year-old atheist, jokingly states: "It's not something I want to talk about on a first date." Apparently, nonreligion is not part of conversation and especially not a way to introduce oneself. According to Joe, a 20-year-old atheist, "It's not something [I say] flat-out that I'm an atheist." Even among friends, Joshua, a 19-year-old atheist, states, concerning nonreligion: "[My friends] don't talk about it, I don't talk about it, it's not that we are avoiding it at all, it's just not really spoken of."

These are commonplace descriptions of everyday life, according to our respondents. Minor variations existed, however. Consider Rick, a 21-year-old atheist: While he has not had an experience wherein he has revealed his nonreligious identity, he still challenges, in careful ways, the idea that discussing nonreligion in everyday interactions is inappropriate. Notice that he both challenges but confirms the non-normative nature of (non) religious discussion. He states:

> Wherever you are it's something I am kind of hesitant to talk about. I will mention that and I will hint at it, but I don't really explicitly say I… [am an atheist] to people. People don't really talk about it, they don't normally ask me.

When at work, the discussion of nonreligious ideas was perceived to be forbidden. According to Greg, a 21-year-old atheist, "I just didn't bring it up at all. I didn't think it would be a wise idea to bring it up [at work]." Though not perceived to be explicitly forbidden, discussion of nonreligious issues in the home was a norm more carefully followed. Samuel, a 45-year-old agnostic, was raised in a somewhat different cultural context as compared to our college-aged respondents; however, his description of being raised in a family where such issues were uncommon rings consistent with younger generations. He states:

> I mean, the whole, you know, you don't discuss politics and you don't discuss religion.... You don't bring it up.... Where I grew up it is something that really doesn't come up much.… It was just never discussed in my house.… [Religion] wasn't really discussed much. It was just like a whole void. That whole thing just didn't happen.

Other respondents confirmed this finding. About nonreligion at home:

"It wasn't discussed at all." *Charles, atheist, 19 years old*
"No we don't avoid discussing it, it's just we haven't discussed it that much." *Joshua, atheist, 19 years old*
"Religion just doesn't come up in our family. So there is no point in really talking about it." *Logan, atheist, 24 years old*

Discussing nonreligious ideas with parents was explicitly avoided. Iconic of this avoidance was Charles, a 19-year-old atheist, who states, "Yeah, [my immediate] family was not really kept in the loop about that." By "that" Charles is referring to his transition from a religious to nonreligious identity. Likewise:

Interviewer: Do your parents know?
Responder: That I'm atheist? No. *Joshua, atheist, 19-year-old male*

The reason was mainly that it served no one to be open and honest about one's nonreligious identity, especially around immediate family members such as parents. Harry, a 19-year-old atheist, summarizes the point nicely:

Interviewer: Have you discussed your transition to nonreligion with your parents?
Responder: Heh [laughs], no actually, it would probably… it probably wouldn't be to anyone's benefit to discuss this with them.

While parents were avoided, more than anyone, grandparents were avoided. They were considered too intolerant and too close to death to bother with such a troubling prospect that their grandchildren were religiously unaffiliated. Greg, a 21-year-old atheist, "I won't bring it up to my grandparents or anything because you know...they need whatever they have." Logan, a 24-year-old atheist, reports that he does not want to interrupt their happiness, as he imagines revealing his identity to them would create unhappiness:

Interviewer: Are there times when you feel that you have to conceal your stance on religion?

Responder: Around my grandparents. I think it just makes them feel happier if they think I still believe.

Sally, a 19-year-old atheist, comparing her contemporary cultural context to the more restrictive context in which her grandmother was raised, suggests that she was raised at a time of intolerance toward nonreligious people and that she holds firm to these beliefs even now. Moreover, because of this fear that her grandmother might find out, this compels Sally to hide her identity from others in the family. She tells us:

Responder: I will never discuss it with my Gram because she is so set in her ways and back in the old days when she was younger atheism wasn't something they really talked about back then so I would probably never discuss it with her. I don't know what her reaction would be and she is just so set in her ways that I don't think I would ever discuss it with her.

Interviewer: Is there anyone else outside of your grandmother that you made sure not to tell?

Responder: One of my good friends is actually my cousin and I don't talk about it because if I plant the seed in her then the whole side of my family would hear about it. They wouldn't be hearing about it from me they would be hearing about it from what she says to them, not necessarily what I want them to hear from me. So I probably wouldn't say anything to her because she would tell everyone, not in a way that I would tell them.

Jeremy, a 19-year-old atheist, suggested that his grandmother's temporal closeness with death influences his decision to hide his nonreligious identity from her. When asked "Is there anyone that you would make an effort not to share it with?" Jeremy states:

Responder: Yea.
Interviewer: Who? I mean you don't have to name names.
Responder: Well I told my grandma I gave up soda for lent and that's an obvious lie.
Interviewer: Why would it be like that with her, that you don't want to share it?
Responder: She's like old and she always thinks she's going to die and so she's all worried about her soul and stuff like that, I don't want to give her anything to think about too much.

All of these interactions are based on imagination. It is how parents *might* respond. How grandparents are *likely* to react. However, few of these interactions come to pass. This is not a coincidence, as many respondents describe measures that they take as precautions in order to completely avoid such exchanges. Phyllis, a 67-year-old atheist, explains that she does not hide her identity from family so much as she does neighbors:

Responder: I don't want to be shunned and spat at... [by neighbors].
Interviewer: Has that happened?
Responder: Oh, of course not! I am not letting it happen! Oh no, you see I don't want to have to go through that. I live there, I live there.... I can be surprisingly good at getting along very well with people of faith, but I need an exit door that I can then go home and relax and be comfortable, but that is my home now. That house is my home now and I want to say hello to my neighbors and get along with the people and not be the one, and not be shunned. I will be shunned frankly.

The acts of concealment—concealing to family or neighbors—have commonalities. Respondents suggest that the main reason for concealment is to remain comfortable around those individuals they interact with frequently.

If Asked, Nonreligious Identity Would Be Revealed

Respondents, with the knowledge that discussion of religion and nonreligion is rare, if not avoided, suggest that if they were explicitly and directly asked in the context of a conversation "Are you a nonbeliever?" then they would open up about their nonreligious identity. Respondents suggest that they are not actively hiding their identity as nonreligious, even though they rarely reveal it. What appears to be a contradiction is actually a subtle strategy designed explicitly to displace the onus of revelation of the nonreligious self in an interactional space away from the identity's owner and onto other members of the interaction.

Dorothy, a 29-year-old atheist, describes the uneasy disconnect between the desire to reveal her identity and her lack of willingness to initiate the interaction. She also, we believe, hints that she secretly desires to be freed from this passive hiding, if only another would compel her to out herself. Dorothy also describes the tension of being simultaneously a silenced and open person. She states:

Responder: What's my religious affiliation? Easy: I don't have one. I don't believe in any religion, but I do find myself falling into their trappings. For example, I've been going to church with my family—they're Catholic—for years. I go for them and I think they sort of, uh, know that I'm not a believer, but we've never really discussed it like formally or whatever. It's a funny thing, not believing, because personally I am proud of it like I've seen through some dense fog everyone else is stuck in. But [pause], but publicly, I sort of keep to myself. I mean, if somebody asked I'd say it in a heartbeat "I am a proud non-believer." But do I bring it up? I mean, not really, not that I wouldn't, [pause] but I just don't, not really. I'm not silenced and I don't feel silenced, it just never really comes up.

Interviewer: Who was the last person you did tell that you were nonreligious?

Responder: I [pause] it's been so long, I can't, uh, seem to remember. I would say something, if asked, though.

Chuck, a 22-year-old agnostic, also describes the feeling that he is hiding without hiding his identity; however, notice that he includes a caveat that he would rather not be the one to raise the issue, in particular, during interactions with members of the Christian faith. Moreover, while he would likely avoid raising the topic, if he were squarely asked, then he would reveal his lack of belief in God. When asked "Are there ever times when you conceal your stance on religion, like where you feel like you have to hide it from other people?" Chuck states:

> I wouldn't say hide, but if I am around a bunch of Christians then I am not going to start talking about how I don't believe in God, just out of manners. If a couple people want to have a conversation then I will, but if I am around Christians and they ask if I believe in God then I will say no, but it's not like I feel pressured into hiding my beliefs.

Chuck's closing words ring consistent with other respondents stating that they either do or do not want to feel, as he states, "pressured into hiding my beliefs." Additionally, notice that he hints at the idea this would only be appropriate in a civil conversation, which might imply that if he reads individuals to be reasonably open-minded during interaction then (and perhaps only then) would he reveal this identity and lack of belief. Greg, a 21-year-old atheist, similarly describes how his willingness to enter into interactions where nonreligion is discussed is contingent on estimating *in situ* the willingness of others to openly consider the matter. There is no more obvious signal that they are willing to consider the matter than when they initiate the conversation. If they do not raise the topic, this appears to be ample evidence that the topic is unwelcome. When asked "Do you try to avoid confrontation with people who are religious?" Greg states: "Um, no. I mean I'm not going to... I'm not going to start questioning people's beliefs who aren't like willing to engage in discussion and who aren't willing to bring it up to begin with."

Sally, a 19-year-old atheist, sees such interactions as akin to interactional chess. Provided the other party in the interaction makes the first move, her actions are merely reaction. The burden of raising the issue is left to the other individuals during conversation. However, notice that Sally unambiguously identifies "close friends" as a suitable audience for her revelation. She states:

Responder: I am not like, closed doors but I am not going to just come out and say it, because you don't just come out with your religion right away but if someone wants to talk about it then I am perfectly fine talking about it... [but] I don't just come right out; I wait until someone says the first thing or asks about it. I won't just come right out and say it. But if they say something about it or ask about it, or if I am really close friends with them, then it's not a big deal I just say it.

Interviewer: When is the most recent time you recall this happening?

Responder: I [pause] I don't really [pause], I don't remember.

Sally, despite her insistence on divulging her identity, cannot recall the last time she did. Given the infrequency of these sorts of events, it is understandable that they may not be fresh in memory. However, the inverse could also be argued: because of the infrequency, one might imagine it would be more memorable and less commonplace (given that if it were commonplace, numerous examples would be available).

Disclosing one's identity is almost exclusively the outcome of being directly asked, meaning that the identity goes ambiguously unidentified without the initiative of a fellow discussant during interactions. Harry, a 19-year-old atheist, is clear about this issue: "I don't really discuss this with many people other than people who ask about it." So is Jeremy, a 19-year-old atheist: "I mean if someone wants to ask me I would tell them." Joshua, another 19-year-old atheist, gets to the crux of the matter, revealing that he would, but does not reveal his identity: "I mean I would have no problem telling my friends, but I just leave it out." Logan, a 24-year-old atheist, does not hide his identity, he states: "I don't really feel like bugging people who are not making it an issue." The boundary drawn between raising the issue oneself and having it raised by another during interaction is significant for nearly every respondent, and clearly stated by Greg, a 21-year-old atheist. When asked "Are there times, to this day, when you conceal your stance on religion?" Greg states "I don't, I just don't bring it up. I mean there is a difference between just not bringing it up and if someone else brings it up."

Even though respondents infrequently report discussing nonreligious ideas or revealing themselves to be nonreligious, some confide in others, but even this is somewhat marred by the "don't ask, don't tell" dynamic our respondents seem to count on. For Jeremy, a 19-year-old atheist, even though he has people to talk to about these issues, he reports that he would rather avoid the topic, while at the same time arguing that he is not holding back his feelings, unless someone else raises the topic. When asked "Did you have people to talk to?" he states:

> Well yea, but I tend to keep things to myself. . . . I mean there were people available I mean I kind of keep things to myself, I didn't hold back my feelings, if someone said something about God I'd be like no thanks or whatever. I'm not one of those people that will announce something like "well guess what?!"

DISCUSSION

Joshua, a 19-year-old atheist, could be any of our respondents. When asked "When was the last time that you talked about [your atheist identity] openly?" he states, "I really don't. I feel if you want to keep your friends, you don't really discuss politics or religion with them, those are two pretty heavy topics that can hurt people."

This half-veiled nonbeliever demonstrates the putative understanding that religion is not a topic of choice in everyday exchanges. This frequent absence of religious discussion allows for the "displaced burden of disclosure," developed independently by multiple atheists, to obscure nonreligious identity. The reluctance of individuals to talk about religion allows nonbelievers to avoid the perceived problems that come with exposure—and they use this resource to their advantage. One issue branches from the *anticipated reaction* that nonreligious respondents imagine will unfold should they reveal their identity. For example, Joshua, a 19-year-old atheist, suggests that he is protecting his family members from emotional harm by keeping his identity hidden. He states, "I try to keep it away from my mom and dad, they wouldn't be happy about it and I mean I love my parents, they are great people, but I just don't see the need to get into that with them and unnecessarily hurt them."

To a religious family, being nonreligious makes their child or sibling an outcast, or at least that is what nonreligious individuals believe will be the likely result of divulging such an identity. However, this hidden identity is not really hidden, for it can be ordered to reveal itself should a friend or family member get wise to the nonreligious person's gambit.

Our respondents utilize a secretive tactic during interactions that is unknowable to others. This ploy is a double-imagined scenario that very few respondents have any first-hand experience with: If asked explicitly about being a nonreligious person in the course of conversation, they insist that they would reveal their identity. Given that discussing religious and nonreligious matters is rare, they imagine an unlikely scenario wherein they will be asked directly. They also imagine their response, as if they would be, in their imaginations, resolute when confronted with the actual question during interaction.

The gamble they make is that no one will ever directly ask. However hypothetical these interactions appear, they assume a modicum of weight as they shape contemporary interactions with family and friends. In effect, concerns about the future, and gambits played based on them, shape the present and in turn shape the future being imagined in what we termed an imagined recursive process.

The practical outcome for individuals on-the-ground is that a nonreligious person can hide without hiding his or her identity. The elaborate imagination employed by respondents suggests that they are not hiding their identity. The practical consequence is that the identity is not divulged and therefore is hidden and hidden again, over and over. However, respondents insist that they are not concealing themselves—and they seal this intrapersonal deal with the promise that if asked, the personal identity is bound to the social identity to reveal and align these two parts of the self.

These multiple ways of "hiding without hiding" feed off of one another; not telling family or friends and the perception that nonreligious individuals will discuss these matters if asked are both allowed to operate because of the fact that religion is not a topic of small talk. Among our respondents, it seems that having a nonreligious family member or friend is something one "discovers" through inquiry rather than something openly discussed or information freely delivered. When it comes to being nonreligious, you have to ask to find out.

CONCLUSION

This chapter begins with an introduction to recursion and its relevance to the social sciences by introducing "imagined recursivity," which is a social psychological conceptualization of recursion. This conceptualization is derived from Cooley's (1902) notion of the looking-glass self. We apply these ideas to a contemporary example: stigma management techniques of the nonreligious.

Cooley's (1902) looking-glass self is a process in which individuals imagine how they appear to others, imagine how others will react to their appearance, and then act according to these imagined perceptions and reactions. Therefore, when an individual enters an interaction, she initiates a recursive process. Individuals perceive themselves through the imagined perception of others in the situation. In effect, their self is looping back on itself via the imagined judgments of others. In this chapter, we introduce a stigma management technique developed by nonreligious individuals that explicitly involves this recursive process.

Nonreligious identity carries with it a weight of stigma. People who affiliate with this identity must find ways to deal with the stigma that, if revealed, it bears. The fact that multiple "Nones" (i.e., religiously unaffiliated individuals) have developed this method independently shows that the displaced burden of disclosure is the preferred stigma management technique among our respondents. Therefore, we believe that this technique must have some substantial benefits to those who employ it. Foremost, if one carefully imagines that "coming out" will be met with ridicule, then there is an obvious desire to avoid such scenarios. Avoiding a situation that will, presumably, result in a fight is usually favored to an actual fight. This technique is incredibly easy to employ because of the recursive process outlined previously. The ease of use allows for this technique to be employed early and often. Also, for many of these Nones, it is much easier to use this technique to conceal their identity than to convince themselves to reveal it. The overriding benefit, of course, is to avoid stigma. With any stigmatized identity there is always the possibility of experiencing unfair, biased, hurtful, and even threatening behavior or treatment upon discovery.

This technique has an intrinsic subset of problems. The first liability is that many of these people are, in an indirect way, lying to themselves and others. Such lying, however, is assuaged by the assurance that they will "fess up" if confronted about religious identity. Second, another liability is that the method being employed is entirely based on speculation. Respondents are only imagining how these interactions might unfold. This imagined scenario is enough to encourage the ongoing use of the displaced burden of disclosure. Several respondents reported, with some sorrow, that while they did not lack someone to fully confide in, they refused to initiate any discussion regarding their stigmatized identity. This, we argue, is one of the premier dangers of displacing the burden of disclosure. Conceivably, and given their increasing numbers, this scenario seems all the more likely: While interacting with others, Nones could be "faking to fakers." In other words, when two nonreligious people interact with one another, neither may be willing to raise the question of religion or nonreligion, which results almost comically in one atheist "faking" to another atheist who is also "faking" back. In the end, neither party learns about the other's hidden identity nor ever develops a relationship that involves discussing this stigmatized identity.

The belief that there will be negative sanctions incurred when another learns that one is nonreligious gives us some understanding of why these situations are avoided. According to our respondents, if family members find out they may disown, be disappointed in, "act weirdly," get into an argument, attempt to convert, think it is a joke, or any number of other negative reactions. Still, the interactional maneuver to avoid disclosure, unless explicitly asked, is largely based on imagined scenarios. This, we believe, is especially significant for Nones. A number of our respondents considered and described themselves as logical and scientifically minded. Further investigation is needed in order to assess how widespread this trait is among nonreligious people; however, this peculiarity warrants mention. Consider Joshua, a 19-year-old atheist, who defines himself by the search for scientific fact and philosophical truth. He is not compelled by belief. Instead, about being nonreligious, Joshua states: "I am seeking the truth, I am not seeking godlessness." There is, if read in a certain light, a shred of irony in the statement. The displaced burden of disclosure, as a method for handling or obviating interactions, is almost entirely based on conjecture—that is, conjecture rather than truth or fact. Nones who adopt this method *believe* that there will be negative consequences in these *imagined* situations of disclosure. Of our self-described hyper-rational respondents, they base their own behavior on *beliefs* about the future rather than hard facts and carefully analyzed personal experience.

It is of particular interest that in order to inject recursion into current social psychological work, we had to "loop back" to one of the original theorists. Cooley's notion of the looking-glass self (1902) has been a hallmark of social psychology for over 100 years. The goal of this chapter is to reinvent Cooley's notion, while at the same time extending it to include imagined recursivity.

REFERENCES

Altemeyer, B. (2004). The decline of organized religion in western civilization. *International Journal for the Psychology of Religion, 14,* 77–89.

Clifford, P. A., Friesen, S., & Jardine, D. (2001). *The ontology of hope: Lessons from a child.* Paper presented at AERA 2001: What We Know and How We Know It, the 82nd Annual Meeting of the American Educational Research Association, Seattle, WA: April 10–14, 2001.

Cocker, J., Major, B., & Steele, C. (1998). Social Stigma. In D. T. Gilbert, S. T. Fiske, & G. Lindzey (Eds.), *Handbook of social psychology* (pp. 504–508). Boston, MA: McGraw-Hill.

Cooley, C. H. (1902). *Human nature and the social order.* New York, NY: Charles Scribner's Sons.

Cragun, R. T., Kosmin, B., Keysar, A., Hammer, J., & Nielsen, M. (2012). On the receiving end: Discrimination toward the non-religious in the United States. *Journal of Contemporary Religion, 27*(1), 105–127.

Dillon, M. (1999). *Catholic identity: Balancing reason, faith, and power.* Cambridge, UK: Cambridge University Press.

Edgell, P., Gerteis, J., & Hartmann, D. (2006). Atheist as "Other": Moral boundaries and cultural membership in American society. *American Sociological Review, 71*, 211–234.

Epstein, C. F. (1988). *Deceptive distinctions: Sex, gender, and the social order.* New Haven, CT: Yale University Press.

Falk, G. (2001). *Stigma: How we treat outsiders.* New York, NY: Prometheus Books.

Gecas, V., & Schwalbe, M. L. (1983). Beyond the looking-glass self: Social structure and efficacy-based self-esteem. *Social Psychology Quarterly, 46*(2), 77–88.

Glaser, B., & Strauss, A. (1967). *The discovery of grounded theory.* Chicago, IL: Aldine.

Goffman, E. (1959). *The presentation of self in everyday life.* London, UK: Penguin.

Goffman, E. (1963). *Stigma: Notes on the management of spoiled identity.* New York, NY: Prentice Hall.

Goffman, E. (1986). *Stigma: Notes on the management of spoiled identity.* New York, NY: Touchstone.

Gray, P. (1994). *Psychology.* New York, NY: Worth Publishers.

Harries-Jones, P. (1995). *A recursive vision: Ecological understandings of Gregory Bateson.* Toronto, ON: University of Toronto Press.

Heatherton, T. F., Kleck, R. E., Hebl, M. R., & Hull, J. G. (2003). *The social psychology of stigma.* New York, NY: The Guilford Press.

Hewitt, J. P. (2000). *Self and society: A symbolic interactionist social psychology.* Needham Heights, MA: Allyn & Bacon.

Hout, M., & Fischer, C. S. (2002). Why more Americans have no religious preference: Politics and generations. *American Journal of Sociology, 67*, 165–190.

Hunsberger, B. E. (1978). The religiosity of college students: Stability and change over years at university. *Journal for the Scientific Study of Religion, 17*, 159–164.

Hunsberger, B. E., & Altemeyer, B. (1997). *Amazing conversions: Why some turn to faith and others abandon religion.* Amherst, NY: Prometheus.

Hunsberger, B. E., & Altemeyer, B. (2006). *Atheists: A groundbreaking study of America's nonbelievers.* Amherst, NY: Prometheus.

Hurh, W. M., & Kim, K. C. (1990). Religious participation of Korean immigrants in the United States. *Journal for the Scientific Study of Religion, 29*, 19–34.

Hurley, S. (2008). The shared circuits model (SCM): How control, mirroring, and simulation can enable imitation, deliberation, and mindreading. *Behavioral and Brain Sciences, 31*, 1–58.

Kosmin, B. A., & Keysar, A. (2009). American Religious Identification Survey (ARIS 2008): Summary Report, March 2009. Retrieved from www.americanreligionsurvey-aris.org

Kroeger, B. (2004). *Passing: When people can't be who they are.* New York, NY: PublicAffairs.

Lamont, M., & Fournier, M. (1992). *Cultivating differences: Symbolic boundaries and the making of inequality.* Chicago, IL: University of Chicago Press.

Levy, S. D. (2009). Becoming recursive: Toward a computational neuroscience account of recursion in language and thought. In H. van der Hulst (Ed.), *Recursion and human language* (pp. 371–392). Berlin, Germany: Mouton De Gruyter.

Link, B. G., & Phelan, J. C. (2001). Conceptualizing stigma. *Annual Review of Sociology,* 27, 363–385.

Mead, G. H. (1934). *Mind, self, and society: From the standpoint of a social behaviorist.* Chicago, IL: University of Chicago Press.

Peek, L. (2005). Becoming Muslim: The development of a religious identity. *Sociology of Religion, 66*(3), 215–242

Scheff, T. J. (2005). Looking-glass self: Goffman as symbolic interactionist. *Symbolic Interaction, 28*(2), 147–166.

Schneider, J. W. (1988). Disability as a moral experience: Epilepsy and self in routine relationships. *Journal of Social Issues, 44,* 63–78.

Serpe, R. T., & Stryker, S. (2011). Interactionist perspective and identity theory. In S. J. Schwartz, K. Luyckx, &V. L. Vignoles (Eds.), *Handbook of identity theory and research: Volume 1, Structures and Procedures* (pp. 225–248). New York, NY: Springer.

Shaffer, L. (2008). Religion as a large-scale justification system: Does the justification hypothesis explain animistic attribution? *Theory & Psychology, 18*(6), 779–799.

Snow, D. A., & Anderson, L. (1987). Identity work among the homeless: The verbal construction and avowal of personal identities. *American Journal of Sociology, 92,* 1336–1371.

Spilka, B., Hood, R. W. Jr., Hunsberger, B., & Gorsuch, R. (2003). *The psychology of religion: An empirical approach.* New York, NY: Guilfold Press.

Stryker, S. (1980). *Symbolic interactionism: A social structural version.* Menlo Park, CA: Benjamin/Cummings.

Swidler, A. (1986). Culture in action: Symbols and strategies. *American Sociological Review, 51,* 273–286.

Sykes, G., & Matza, D. (1957). Techniques of neutralization: A theory of delinquency. *American Sociological Review, 22,* 664–670.

Thompson, W. E., & Harred, J. L. (1992). Topless dancers: Managing stigma in a deviant occupation. *Deviant Behavior, 13,* 291–311.

Vaughan, D. (1996). *The Challenger launch decision: Risky technology, culture, and deviance at NASA.* Chicago, IL: University Of Chicago Press.

Warner, W. L. (1993). Work in progress toward a new paradigm for the sociological study of religion in the United States. *American Journal of Sociology, 98,* 1044–1093.

Yeung, K.-T. & Martin, J. L. (2003). The looking glass self: An empirical test and elaboration. *Social Forces, 81*(3), 843–879.

LIST OF INTRODUCTORY TEXTS

Albrecht, S. L., Chadwick, B. A., & Jacobson, C. K. (1986). *Social psychology* (2nd ed.). Upper Saddle River, NJ: Prentice Hall College Division.

Arlsdale, J. Z. (2003). *Trends in social psychology*. Huntington, NY: Nova Science Publishers.

Aronson, E. (1999). *The social animal* (8th ed.). New York, NY: W. H. Freeman & Company.

Baron, R. A., & Byrne, D. (1977). *Social psychology: Understanding human interaction* (2nd ed.). Boston, MA: Allyn & Bacon.

Baron, R. A., Graziano, W. G., & Stangor, C. (1991). *Social psychology* (2nd ed.). Independence, KY: Delmar.

Berkowitz, L. (1986). *A survey of social psychology* (3rd ed.). New York, NY: Holt Rinehart & Winston.

Brown, R. (1985). *Social psychology* (2nd ed.). New York, NY: Free Press.

Callan, V. J. (1986). *Social psychology*. San Diego, CA: Harcourt, Brace, Jovanovich.

Chadee, D., & Young, J. (2006). *Current themes in social psychology* (illustrated edition.). Kingston, Jamaica: University of West Indies Press.

Charon, J. M. (1989). *Symbolic interactionism: An introduction, an interpretation, an integration* (3rd ed.). London, UK: Longman Higher Education.

Cvetkovich, G. (1984). *Social psychology: Contemporary perspectives on people*. Holt Rinehart and Winston.

Durkin, K. (1995). *Developmental social psychology: From infancy to old age*. London, UK: Blackwell.

Eiser, J. R. (1986). *Social psychology: Attitudes, cognition and social behaviour* (rev. updated ed.). Cambridge, UK: Cambridge University Press.

Gray, P. (1993). *Psychology 2/E: Subj* (2nd ed.). Institute of Electrical & Electronics Engineering.

Hamilton, P., & Thompson, K. (2002). *The uses of sociology*. New York, NY: Wiley-Blackwell.

Jahoda, G. (2007). *A history of social psychology: From the eighteenth-century enlightenment to the second world war* (1st ed.). Cambridge, UK: Cambridge University Press.

Kalat, J. W. (2004). *Introduction to psychology* (7th ed.). London, UK: Wadsworth Publishing.

Kassin, S. (2003). *Essentials of psychology*. Englewood Cliffs, NJ: Prentice Hall.

Kendler, H. H. (1977). *Basic psychology* (Brief version). Menlo Park, CA: W.A. Benjamin.

Kosslyn, S. M., & Rosenberg, R. S. (2000). *Psychology: the brain, the person, the world*. Boston, MA: Allyn & Bacon.

Kowalski, R. M., & Westen, D. (2004). *Psychology* (4th ed.). New York, NY: Wiley.

Kremer, J., Sheehy, N., Muldoon, O., Reilly, J., & Trew, K. (2003). *Applying social psychology*. London, UK: Palgrave Macmillan.

Lange, P. A. V. (2006). *Bridging social psychology: Benefits of transdisciplinary approaches*. Mahwah, NJ: Lawrence Erlbaum Associates.

Morris, C. G., & Maisto, A. A. (2001). *Psychology: An introduction* (11th ed.). Englewood Cliffs, NJ: Prentice Hall.

Moskowitz, G. B. (2001). *Cognitive social psychology: The Princeton Symposium on the Legacy and Future of Social Cognition.* Mahwah, NJ: Lawrence Erlbaum.

Myers, D. G. (1998). *Psychology* (5th ed.). New York, NY: Worth Publishing.

Nairne, J. S. (1999). *Psychology: The adaptive mind* (2nd ed.). Belmont, CA: Wadsworth.

Passer, M. W., & Smith, R. E. (2001). *Psychology: Frontiers and applications.* New York, NY: McGraw-Hill College.

Passer, M. W., & Smith, R. E. (2003). *Psychology: The science of mind and behavior* (2nd ed.). New York, NY: McGraw-Hill.

Plotnik, R. (2005). *Introduction to psychology* (7th ed.). Belmont, CA: Wadsworth.

Rathus, S. A. (2000). *Essentials of psychology* (6th ed.). New York, NY: Wadsworth.

Rubin, Z., & McNeil, E. D. (1985). *Psychology: Being human* (4th ed.). New York, NY: HarperCollins College Div.

Schaller, M., Simpson, J. A., & Kenrick, D. T. (2006). *Evolution and social psychology.* New York, NY: Psychology Press.

Sdorow, L. R. C. A. (2001). *Psychology* (5th ed.). New York, NY: McGraw-Hill.

Smith, E. E., & Kosslyn, S. M. (2009). *Cognitive psychology: Mind and brain.* Englewood Cliffs, NJ: Prentice Hall.

Stolley, K. S. (2005). *The basics of sociology.* Westport, CT: Greenwood Press.

Taylor, S. (1999). *Contemporary sociology.* London, UK: Palgrave Macmillan.

Turner, P. B. S. (1999). *Classical sociology.* Thousand Oaks, CA: Sage.

Wade, C., & Tavris, C. (1999). *Psychology* (6th ed.). Upper Saddle River, NJ: Pearson Education.

Zimbardo, P. G., & Gerrig, R. J. (1998). *Psychology and life (with SuperSite and MindMatters CD-ROM)* (15th ed.). Boston, MA: Addison Wesley Longman.

CHAPTER 5

UNDERSTANDING OTHERS WITHOUT A WORD

Articulating the Shared Circuits Model with Semiotic-Cultural Constructivist Psychology

Danilo Silva Guimarães
Institute of Psychology–University of Sao Paulo

Andre Mascioli Cravo
Centre for Mathematics, Computation and Cognition–Federal University of ABC

In the present chapter, we attempt to articulate the theory of shared circuits model (SCM) developed by the philosopher Susan Hurley (1954–2007) with the semiotic-cultural constructivism in psychology (cf. Simão, 2003, 2005, 2010). Through the proposed theoretical articulation we aim to develop a background for further investigations in the field of an integrative approach on phylogenesis, ontogenesis, and social relationships. Even though the questioning on the embracement between biological and social histories has been widely debated in psychology, theoretical

Cultural Psychology of Recursive Processes, pages 143–161

conceptions on this relationship remain polemic and fragmented. We propose that recursive epistemology (cf. Bateson, 1972) is convergent with the present integrative effort, since the ecology of the mind addresses dynamic, integrative, and multilevel understandings of what happens in the boundary between the inside and the outside of an organism (cf. Bateson, 1972; Ingold, 2000).

Recursivity is being understood here as a property of interdependent and irreducible systems. Applied to the relation between an organism and its environment, recursivity implies that the organism needs to change in order to apprehend information from the world, at the same time that adaptive changes allow the maintenance of the organism in the environment. However, the environment also changes due to actions of the organism. The recursive process is, then, bidirectional and of growing complexity.

From the psychological approach of the semiotic-cultural constructivism, the relation between a person and the environment is also viewed as recursive: The semiotic-cultural constructivism is a "guiding, temporary and flexible character of an instrument for qualitative research [that] focuses especially on the process of individual development, in which I–other interactions unfolding from, as well as forming, the cultural space have a prime role" (Simão, 2003, p. 550). In this process, the interlocutor's quest for mutual understanding and sharing of their experiential meanings brings about a movement of decentration, from which some novelties can emerge in their comprehension. In such developmental process, hierarchical levels can emerge and reorganize the interactive dynamics:

> By semiotic-cultural constructivism, I mean the contemporary theoretical methodological perspective in psychology developed mainly around the ideas of Lev Vygotsky, Mikhail Bakhtin, George Herbert Mead, Pierre Janet, Jean Piaget and, more recently, Ernst Boesch and Jaan Valsiner, amongst others. (Simão, 2005, p. 550)

Another important predecessor of the developmental ideas concerning the semiotic-cultural constructivism in psychology is James Mark Baldwin (cf. Simão, 2008). We will revisit some of Baldwin ideas about child mental development, in light of recent findings about imitation, especially relating imitation with perception, action, and mindreading[1] We will draw links between Baldwinian pioneer ideas on child mental development with the theory of shared circuit model proposed in the core of contemporary cognitive studies. At the end of the text we will explore an articulation between the Shared Circuit Model (Hurley, 2006, 2008) and the affective schemata recently developed by Valsiner (cf. 2001, 2007a) as a strong reference to understand the semiotic-cultural guidance of feelings.

AN INTEGRATIVE APPROACH TO UNDERSTANDING THE EMERGENCE OF CULTURAL MEANINGS

Language is usually considered as the main device of human sociability. The possibility of sharing understandings and giving consistency through semiotic representations allows the construction of a set of culturally structured meanings (cf. Boesch, 1991; Duran, 2004; Valsiner, 2001, 2008). The bodily and preverbal interactions are normally thought as secondary when compared to the multiple forms of meditational use of signs and symbols. Embodied expressive interactions are also considered more primitive than others. Fogel, Koeyer, Bellagamba, and Bell (2002) proposed that the dialogical self—a theory about how subjectivity is constructed through the internalization of social voices (cf. Hermans, Kempen, & van Loon, 1992)—develops in infancy from nonverbal and embodied conflicts that become predominantly narrative after eighteen months.

Semiotic-cultural psychology maintains that the social representation is a construction that organizes the experience (cf. Duran, 2004). To represent, through semiotic devices, is an intersubjective symbolic action that comprises of positioning something to someone (cf. Marková, 2000, 2006). The flow of feelings is constrained by semiotic devices, which gives labels to nonverbal experiences, allowing generalization and also guiding affective-cognitive trajectories (cf. Josephs, 2000; Valsiner, 2001, 2007a, 2007b). Nevertheless, human symbolic trajectories are personally centered in an organic structure in which diverse self-regulatory mechanisms coexist and are integrated (Oppenheimer, 1991).

On the other hand, contemporary anthropology on Amerindian culture emphasizes processes of body fabrication as the main investment on human upbringing for a great number of autochthon communities from America (cf. Lagrou, 2007; Seeger, da Matta, & Viveiros de Castro, 1979; Taylor, 1984/1996; Viveiros de Castro, 2002/2006). Integrative concerns permeate some anthropological researches about how the body is "constructed" to regulate social interactions. The anthropologist Tim Ingold (2000) argues in favor of the construction of an integrative framework among phenomenological, cognitive, and anthropological studies in order to understand Amerindian ways of life. From phenomenology he takes the notion of immanency, an inclusive approach of being-in-the-world. Semiotic-cultural constructivism, in psychology, is also challenged to address an integrative approach among phenomenological, cognitive, and anthropological perspectives (cf. Guimarães, 2011). Addressing this path, the description of the exchange between the organisms and their environment through a set of recursive layers (Hurley, 2006, 2008) can be integrated with the description of self regulatory mechanisms in the form of recursive levels of semiotic organization, designed by Valsiner (cf. 2007a, p. 312). As these propositions seem to approach different regulatory systems that coexists in organisms, we consider it worthwhile to attempt an articulation between them.

SEMIOTIC-CULTURAL CONSTRUCTIVIST ROOTS FOR CONCEIVING EMBODIED MEDIATED PROCESS

Oppenheimer (1991) argues in favor of a scientific research program that includes contextualism and action theory without neglecting the organismic nature of the object of study. His organismic paradigm asserts that the organic development is a dialectic process that involves thesis, antithesis, and synthesis. He proposed the articulation of different systemic levels starting from the physiology of the organism to its macrossocial positioning, advancing in the path inaugurated by James Mark Baldwin.

Fifty years after Darwin's *On the Origins of Species* (1859/2006), Baldwin (1896) proposed the existence of individual adaptation processes that makes the maintenance of functional characteristics owned by an organism feasible. The organism is able to retain not just phylogenetically hereditary aspects, but also aspects learned throughout its life. It means that the organism can address personally constructed knowledge to future generations, inserting novelties in the path of the cultural history. This phenomenon named "organic selection" (Baldwin, 1896) is called the "Baldwin Effect" within the field of evolutionary studies and can be thought of as a precursor of the Vygotskian proposition of interdependency among phylogenesis, ontogenesis, and sociogenesis (Vygotsky & Luria, 1996). Therefore, natural selection does not mechanically determine evolution; in contrast, according to Baldwin, "evolution is . . . not more biological than psychological" (1896, p. 547)—that is, evolutionary novelties that emerge in the organic history can stabilize with the production of hereditary modifications:

> The organism experiences its environment in an interactive and continuous process, adjusting and changing itself, letting impacts in its epigenetic system, in its genome as well as in its environment. These impacts are left to the subsequent generations. So, there is not any separation between development and evolution. The organism takes an active part in the process of its own development, as well as in the evolution of its ecological community. (Almeida & Falcão, 2008, p. 529)

Including psychological factors in a theory of evolution requires a subtle reconceptualization of organisms, conceiving them as actives upon their environment, which implies that organic variations are not only selected by chance. Complementing the negative postulate from Darwinian "natural selection" theory, which focuses on the consequences for those individuals that do not have the necessary characteristics to survive, organic selection states a positive agency: The organism that is able to learn can transform itself and its environment to create conditions for its own adaptation and for the inclusion of its social group. The notion of learning conceived by Baldwin implies an epigenetic process in which habits are constructed and "broken up" in the ontogenetic dynamics of assimilations and accommodations.

Baldwin (1906a) noticed that among all living beings capable of learning, humans possess the fewest congenital behaviors but are able to learn the most during a lifetime. Human beings, with fewer instincts (fixed-action patterns in response to the environment), are capable of addressing the environment in multiple ways. Consequently, a desirable quest for psychology is to explain the emergence of novelties in this linkage between cognition and volition. To understand the psychological differentiation process, wherein an organism as a whole develops in a singular way, Baldwin proposed some ontogenetic stages in which affect, cognition, perception, and action are imbricate:

> Without insisting on the details of this sketch—intended at this point for no more than a sketch—certain great epochs of functional differentiation may be clearly seen. First, the epoch of the rudimentary sense processes, the pleasure and pain process, and simple motor adaptation, called for convenience the "affective epoch": second, the epoch of presentation, memory, imitation, defensive action, instinct, which passes by gradations into, third, the epoch of complex presentation, complex motor co-ordination, of conquest, of offensive action, and rudimentary volition. These, the second and third together, I should characterize, on the side of consciousness, as the "epoch of objective reference." And fourth, the epoch of thought, reflection, self-assertion, social organization, union of forces, co-operation; the "epoch of subjective reference," which, in human history, merges into the "social and ethical epoch." (Baldwin, 1906a, p. 15)

Organic selection is remarked by some identifiable stages of the ontogenetic process addressing the social environment. These general ideas about human development characterize Baldwin's theory as an inherently cultural approach to the mental development that precludes partial/monological understandings to psychological functions and processes. That is, human development cannot be studied fragmenting cognition from volition or perception from action: On the contrary, it has to be studied articulating all these aspects.

DIVERSITY AND THE ROLE OF IMITATION FOR ORGANIC SELECTION

Baldwin's theory operates with the coexistence of two dimensions that allow development: (1) reactive organic variability and (2) the possibility of repeating or control these reactions. These two dimensions cooperate in what he defined as imitation, a process that permeates the biological-phylogenetic and the social-psychological levels of an organism's life. Thus, imitation is understood as a circular reaction, a recursive action that is addressed to repeat its own stimulus (Baldwin, 1894). In their early life, children physiologically generate a variety of muscular discharges. This variability happens in part accidentally and constitutes an embodied repertoire of actions. Additionally, there is also a biological tendency to reproduce muscular contractions, as a reaction that sustains the copy. Progressively the child starts to be able to inhibit some of them, reproducing just those that are satisfactory, or, in the Baldwinian sense, those that cause pleasure or are useful.

Nevertheless, Baldwin (1896, 1906a) proposed the existence of lower and higher circular reaction, which emerges during the organism's development from simple imitation to persistent imitation. The latter is characterized as an active and selective effort by the child to imitate specific actions. This effort can be linked with the sensitive experience of the child as pleasure or pain avoidance. Additionally, the gestures can also be recovered from memorized repertoire of past experiences. Present and memorized stimuli can be repeated and possibly accumulated into habitual forms of expression by the organism. However, the organism also needs to handle with the multiplicity of paths to follow: There are many possible actions and it has to select one as an answer. The notion of volition is used to describe a developmental acquisition—that is, the possibility to reproduce some actions while inhibiting others. It is worth emphasizing that, contrary to common sense, volition in Baldwin's view depends on the capacity of inhibiting actions, breaking up previous instinctive habits.

Baldwin (1906a) finds evidence of volition in children's persistent imitations, which are not being explicitly suggested to them. Persistent imitation is a precursor of adults' justifications for actions as simply because they like to. However, what adults like to do is grounded in the circumstances of their specific childhood experiences and environment. What is arranged in an organism's memory is, primarily, the result of its material experience and caregiver's suggestions and constitutes its belief of what is existent and what is reasonable in the environment (Baldwin, 1896). Of course, in the ontogenesis of an organism there is always a possibility of rectification of personal concepts. Deviant experiences are accommodated in the basis of growing consciousness about the relationship of elements in a reasonable world. Nevertheless, the structural solidification of ongoing habits strongly depends on the organism conviviality, which recursively channelizes embodied actions and semiotic conceptions.

Finally, Baldwin (1896) states that some links between the copied object and the original object can be lost in consciousness and some actions become automatic. Baldwin proposed the "law of Habit" to understand this phenomenon:

> Physiologically, habit means readiness for function, produced by previous exercise of the function. Anatomically, it means the arrangement of elements more suitably for a function, in consequence of former modifications of arrangement through that function. Psychologically, it means loss of oversight, diffusion of attention, subsiding consciousness. (Baldwin, 1906a, p. 277)

When a habit is settled down, the organisms finds themselves in poor condition to recover past links for their contemporary procedures. This phenomenon allows us to think about the distinction between contemporary personal justifications for actions and other narratives on it. There is a gap between the person in the present and itself in the past, between the personal perspective and the perspective of the other. These gaps are in the core of recursive process which happens in the autopoiesis of the organisms as so as in their open-endedness (Valsiner, 1998) relations with otherness.

LEARNING BY IMITATION: THE "MAKE-BELIEVE" PLAY

The use of inner simulation for knowledge construction about the world and otherness was described by Baldwin (cf. 1906b) in his discussion on a childhood imitative interplay called *sembling* or make-believe (cf. Valsiner, 2000, 2008). The notion of sembling was explored by Baldwin as a fictional, experiential, and selective procedure in the genesis of reflection. As it is impossible to perfectly imitate the other, the "sembler" needs to continuously compare his or her actions with the action of the other, leading him or her to apprehend or imagine, from the schema of a set of actions, the involved aims before the otherness active chain is concluded.

Valsiner (2008) explores the notion of *sembling* emphasizing "a never-ending forward oriented construction cycle where established schemas lead to new created roles for new objects of exploration, while the latter lead to establishment of ever new schemas" (p. 61). Controlled imagination involves knowledge construction aimed at reducing tension at the crossroads of fictional/creative freedom and the effort of corresponding to the object; that is, someone can recursively control her own behavioral variations if she intends to act in a similar way as the other is acting. It allows the person to grasp the intentions of the other, the implications of some actions, and makes viable the emergence of new potentials for further explorations.

The internalization of affective-cognitive schemas for actions also allows the person to imagine the consequences. The contrast between the external reference and the inner simulation is remarked by Valsiner's (2007b) proposition on the subjectively lived duality "as-IS," which refers to the perception of the world "out there," and "as-IF" the inner imagined possibilities of being. This duality addresses human actions from the actual condition to a desired future.

The cultural psychologist Ernst Boesch (1991) also operates with a similar distinction on the basis of the perception-imagination duality. On the one hand, he calls "is value" the objective construction of the environment as separate from the self. On the other hand, the imaginative level would be configured by a subject as a "should value." This means that through the human action upon the world, someone can transform it and/or transform his personal expectancies, constructing a developmental, and at the same time personal and contextual, history. For instance, when a child proposes a new rule to play soccer in the street with his friends, such as proposing a size for smaller goalposts and a distance between the goalposts of each team, he would have an expectancy of adequate the official rules to his environmental condition in order to increase playability. His proposal can be agreed with by his friends and can be settled down as the rule for playing soccer in that community.

In sum, while acting in and on the world, a person is able to estimate self dispositions and abilities to act in many directions face to an environmental configuration. The person is, in many cases, not aware of his own potential for acting. According to Boesch (1991, cf. p. 56), imagination is a process that corresponds to the planning of an action. The symbolic action, as defined by him, departs from an imagination about what someone intends to achieve—SHOULD value—from what he or she perceives the environment in the present—IS value. Additionally, lived experiences evoke models and anticipations that are embodied in different levels of personal perception of the environment.

THE COGNITIVE BASIS OF IMITATION

Although imitation has been widely discussed in cognitive sciences, mainly in developmental studies (see Meltzoff, 2002), the discovery of "mirror neurons" can be considered as a dividing marker in the field. In a series of classical experiments measuring neural activity in monkeys' brains, it was discovered that a particular set of neurons was activated during the execution of purposeful, goal-related hand actions, such as grasping, holding, or manipulating objects, and also when the monkey observed similar hand actions performed by another individual (di Pellegrino, Fadiga, Fogassi, Gallese, & Rizzolatti, 1992; Gallese, Fadiga, Fogassi, & Rizzolatti, 1996; Rizzolatti, Fadiga, Gallese, & Fogassi, 1996). These neurons are located in the premotor cortex, an area originally thought to be responsible for preparation, simulation, and selection of motor functions.

This privileged localization suggests that mirror neurons have facilitated access to the motor cortex, but not directly. A direct connection would imply that the organisms would be constantly imitating all movements they see, while most adults easily inhibit imitation impulses. However, human patients with adult "imitation syndrome" mimic involuntarily (Frith, 1992). They persistently copy the experimenter's gestures, even when not asked to and when these are socially awkward. Still, humans are more susceptible to others' actions than we imagine, often referred to as the chameleon effect (Chartrand & Bargh, 1999). For example, normal adult subjects who interact with someone who constantly rubs his own foot end up rubbing their own foot significantly more (Chartrand & Bargh, 1999).

Therefore, contrary to an intuitive view, imitation seems to be more automatic than we think, although that does not mean that it is a low level or undemanding behavior. In fact, there is still rather large discussion if true imitation exists outside the human species (see Hurley, 2008 for a review). As stated by Susan Hurley: "True imitation requires that a novel action is learned by observing another do it, and, in addition to novelty, requires an instrumental or means/ends structure: you copy the others' means of achieving her goal, not just her goal, or just her movements" (2008, p. 12). For her cognitive approach, the necessity of imitation requiring the copy of means and movement is important since most animals appear to be able to copy only one of these two. In neither case do other animals learn about the intentional means/end structure of the observed action. This comparison is remarkably striking when we compare imitation in chimps and human children. Children have been observed to overimitate, or to reproduce an adult's obviously irrelevant actions, in several different contexts, even in situations where chimpanzees correctly ignored the unnecessary steps (Lyons, Young, & Keil, 2007).

In children, imitation is fairly automatic, suggesting that the inhibition of this behavior comes only with further development. More than merely avoiding awkward social situations, this inhibition may allow more complex behaviors. Consider a person seeing another one act. In the absence of inhibition, the action of the other person would lead to similar (although not necessarily identical) action of the observer. Nevertheless, in the presence of inhibition, the observing of the other's act can still prime the subject's behavior, providing the observer with simulated information about the intentional character of the observed act. In other words, this inhibited simulation can provide the basic information for understanding of others' intentions.

Victor Gallese, one of the researchers involved in the recordings of mirror neurons, has suggested that action understanding heavily relies on a neural mechanism that matches the observed behavior with the one executed. According to his view, "when I observe other acting individuals I can immediately recognize them as goal-directed agents like me, because the very same neural substrate is activated as when I myself am bound to achieve the same goal by acting" (Gallese, 2001, p. 41).

Some of these points can be easily related to Baldwin's propositions discussed before: (1) imitations can be inhibited and the healthy organism is able to select which gestures to imitate; (2) human beings, and especially children, can easily mimic both means and ends of an observed action; (3) we may not be aware of the motives of many of our actions; we simply do it because of an instinctual tendency to imitate the other or because of habit formation in the basis of previous imitative references; (4) the existence of inhibited simulation at the same time that allows the organism to empathically understand the other implies an interdependent distance between self and other. In other words, there is a setting of physiological mediations between self and other, and these mediations are co-regulated by the experience of the singular organism in different levels.

The propositions above lead us to suppose that volition and affective-cultural references are substantively inscribed in the body of human organisms. In this sense, we are working with the notion of cognition and feelings considering that there is continuity between both. According to Stern (cf. Josephs, 2000), feelings are in the boundary between unconscious activity and planning, in the sense that cognition is the possibility to distinguish feelings in an affective nebulous background. On the one hand, cognition is commonly used to refer to reflected oriented actions and decision making, and, on the other, prepersonal models constructed by contemporary cognitive research are being used to demonstrate that many of the planning strategies are developed from a prereflexive and preverbal embodied functioning.

ACTIVE SIMULATION AND INHIBITION OF ACTIONS: INCREASING COMPLEXITY IN IMITATIVE PROCESS

In the shared circuits model (Hurley, 2008), a five-layer model is developed with the starting point being dynamic online motor control. Onto this are added control functions of prediction, mirroring, simulation of mirroring, and monitored inhibition of motor output layer by layer. As Hurley summarizes her own model:

> The shared circuits hypothesis provides a unified subpersonal architecture for control, imitation, and simulation.... The feedback effects of certain motor outputs, such as visual inputs that result from certain movements, are predicted via simulation; such prediction can benefit instrumental control. Mirroring reverses this predictive simulation, so that observation of movements of a certain kind by another induces in the observer motor output that would typically cause such movements. While copying can be beneficial, the capacity to inhibit actual copying is also adaptive. When overt copying is inhibited, the

> causes of the observed behavior are nevertheless simulated, enabling action understanding. The progression is from simulation of effects through mirroring to simulation of causes. (Hurley, 2006, p. 37)

Although by no means all the complexity of the SCM will be addressed in the present chapter, the importance that is given to motor inhibition deserves attention. In the SCM, a monitored output inhibition is essential so that several simulations can be done without performing actual actions. This simulation is important not only for imitation but also to differentiate between actual and possible outcomes. For example, before deciding the best action to be taken, one can simulate several possible couplings of action-outcomes.

However, it is by combining imitation mechanisms with a monitored output inhibition that the most interesting possibilities arise. For instance, imagine seeing someone perform an action. This observation should activate a similar action on you. However, you can inhibit the motor output of this action if copying this behavior is not beneficial at the moment. Even though you are not explicitly copying that action, you can still simulate it mentally, thus helping you to understand the action and intentions of the person you are observing. Notice that in this case self and other share a common informational space. An observed action, even if not imitated, can prime the observer's future actions, while also leading to action understanding and mindreading, in the sense we previously discussed.

One of the most interesting aspects of the SCM is that the shared processing by self and by others is a special aspect of shared processing of action and perception in motor control. Moreover, as Hurley (2008) states:

> These shared resources are prior to self/other and actual/possible distinctions that provide information for action understanding and instrumental deliberation. The shared processing of action and perception in dynamic control is preserved when an actual/possible distinction is overlaid via inhibition of overt action. Similarly, the shared processes of action and perceiving others' action are preserved when a self/other distinction is overlaid via inhibition of overt copying. (p. 21)

Once more we face a rather counterintuitive view: Instead of having self and other as separated representations that are integrated during development, the SCM suggests that we start off with these representations sharing a common space. Even the later development of motor inhibition does not totally separate the common coding between self and other. Therefore, empathy, mindreading, and action understanding can be thought of as not based exclusively on the high-level behaviors such as language. On the contrary, the possibility of an intrinsically shared informational space between self and other suggest that the effects that others have on our behavior emerge from simple process.

REVIEWING THE "PHYSIOLOGICAL LEVEL" OF THE AFFECTIVE SCHEMATA

The SCM enlarges the field of cultural phenomena analysis since it describes how human sociocognitive skills can emerge from basic adaptive feedback controls in the relation of the organism with the environment. Feedback controls, understood as "cyclical and dynamic process" (p. 12) (cf. Hurley, 2008), can produce recursive internal and external loops. On the one hand, according to Werner's orthogenetic principle, the embodied organism is poised to increase internal complexity by organizing feedback hierarchically (cf. Valsiner, 2001). On the other hand, Hurley (2008) proposes that increasingly complex instrumental layers work interdependently and concomitantly in self-regulation. In the SCM, the starting layers allow the organism to improve its control over the environment through simple feed forward simulation of effects.

These process can be linked with the physiological level 0, discussed by Valsiner (2001, 2007a, 2007b) in his hierarchical model of human semiotically mediated affective fields. According to Valsiner (2001), the level 0 corresponds to "the universal—for all animal kingdom—physiological anticipation about the immediate next future event in life. Based on that level, the organism can develop generalized, non-mediated <<feeling tone>>" (p. 163). Nevertheless, Hurley physiological-cognitive model unfolds this level in the direction of embodied mediated nonverbal symbolic cultural layers. That is, according to her, the subsequent layer focuses on the imitation of other's similar acts or evoking objects (Hurley, 2008).

It entails a significant difference in the way as different organisms construct their own anticipations about the immediate next future, based on its convivial particularities. For instance, once intersubjectivity occurs in situated settings, primitive mediation related to visual, nonverbal emulation implies expectancies on the behavior of the other. This mediation is constructed on the basis of the imitation of others symbolic actions.

The SCM describes two more layers. The fourth concerns implications of the organic inhibition of some imitations and actions (Hurley, 2008). The presence of an internal inhibition of imitation bridges a gap between self and others and is hypothesized as the subpersonal background of this distinction. This physiological process allows the organism to become able to also distinguish between the actual and the possible, distancing itself from the present situation. Baldwin (1906a) asserts that the capacity of inhibiting some imitations and to persist in other imitations is an evidence of volition;

one becomes able to show that he is able to select how to behave according to his interests and/or desires.

While the notion of volition is central in Baldwin's theorization (cf. Valsiner, 2000), addressing his theory to psychological development, SCM focuses on subpersonal understanding of intersubjectivity. Hurley (2008) proposes that both subpersonal and personal levels are interdependent. Finally, in Layer 5, the organism can simulate other possibilities of action not only for its own actions, but also for actions of others. This final layer can allow, for example, mindreading, since now the organism can, via inhibited simulations, understand why the other chose that specific action.

At this point language can also start to meddle with the organic mediation. Although Susan Hurley was very careful with the addition of language in the SCM, she does recognize that it can help to organize simulative mechanisms face to different life experiences in a progressive multiple social context:

> Language can build on SCM's foundational actual/possible and self/other distinctions to enable interpretative understanding of multiple others with multiple alternatives and varying beliefs. SCM hypothesizes that mindreading has practical foundations, in simulative mirroring of means/ends relations, but allows mature mindreading with all the bells and whistles (including understanding false beliefs) requires both simulation and language-based theorizing. (Hurley, 2008, p. 18)

For our purposes of exploring a path to understand the integration between phylogenesis and sociogenesis, SCM can be a valuable model to explore more accurately the preverbal level of affective/cognitive experience. A relevant aspect of this model is to remark that the physiological level (prepersonal) is not unidirectional, but possibly organized in different layers of recursive interaction with the environment. Another point is that this experience does not work in the same way for the whole animal kingdom, but each organism evolved with different circular mechanisms to deal with the exteriority. Besides, the neuronal adaptation, in human beings, varies according to intersubjective embodied experiences with the respective group of upbringing. Finally, the body can be understood, from SCM, not as a purely biological entity, but as an intersubjective entity that mediates, without the necessary encoding through signs, the interdependency between internal and external loops.

SOME WORDS ON THE RECURSIVE NATURE OF IMITATION

From what we discussed above, imitation can be thought as a recursive process of reproducing another's or one's own movement. This process can happen in different levels of awareness. Imitating the other in the environment is one possibility of reacting to it that can be evaluated from a developmental perspective, as Baldwin proposed. Imitation is closely connected with perception and imagination once that the personal imitative action inherently carries some supposition about the movement of the other.

The investigation on imitation, articulating the semiotic-cultural constructivist (SCC) framework with the cognitive shared circuit model (SCM) challenges us to reconcile the atomist view of the organism functioning in isolation and the holist, historically oriented comprehension of the interchanges between an organism and its environment.

The atomist approach works with a strict separation of knowledge entities and searches for causality or correspondence between findings in each entity; on the other hand, the holist approach tries to work upon the inclusive separation of perceptive elements, considering not just the distinction but the boundary between the entities. Efforts to overcome this dichotomy are not new to research but continue to disquiet contemporary scholars (cf. Valsiner, 2001; Diriwhachter & Valsiner, 2008). This overcoming echoes different problems identified by physiological and social frameworks associating human and natural sciences, specifically nomothetic and idiographic studies in psychology (cf. Salvatore & Valsiner, 2009).

The approach on recursivity presented here assumes that there are at least three relational entities involved in the imitative process: (1) the person that acts; (2) the other that imitates; and (3) the environment, in which the imitation occurs. Dialogical methodology proposes to take into consideration that I and non-I (the other and the environment) comes to existence together as the relation between figure and ground, in such a way that in composes a tensional boundary between both (Herbst, 1995). It leads to a transformation of the open-ended systems through the semiotic/symbolic mediations, as a way of reducing the tension emerged by the relation.

Such dialogical conception of interdependent entities supposes a distinction between self, other, and environment as differentiated systems with autonomous proprieties, but these systems are also open to each other, allowing mutual—or recursive—influence.

FINAL CONSIDERATIONS: RECURSIVITY IN THE BOUNDARY OF SYMBOLIC ACTION AND PERCEPTION

We departed from Baldwin's investigative roots articulating the biological and the social in the organism ontogenesis, emphasizing that the notion of organic selection demands integrative efforts of structural and historical-processual frameworks. Despite bringing many advances for natural and social sciences of his time, Baldwin was limited by technical and conceptual devices from biology, anthropology, and sociology of that period (Baldwin, 1930). Nevertheless, as we showed here, ideas he addressed are being developed in contemporary cognitive studies on intersubjective relationships, mainly on the process of imitation. These studies have shown that many subpersonal layers coexist in the organism and structure its relationship with the environment and others.

In such a way, our body seems to be inherently addressed to intersubjective experiences. We are intense imitators, especially when young, but still remarkably affected by the action of others when adults. The prepersonal layers discussed by Hurley suggest that an increasing capacity to understand and simulate the action of others can happen even without the use of encoding through signs. Because of constant exposure to an environment that is culturally organized by particular persons, the body is biochemically transformed; some neuronal patterns are intensified as some others expire.

Thus, a major contribution of SCM to semiotic-cultural constructivism was to show how imitation, mindreading, and priming emerge from relatively simple mechanisms. Certain schools of psychology and sociology understand that social behaviors and cultural learning are intrinsically thought to be related to language and other "high-level" functions. However, the contrary view can also be argued. We might be social organisms even before any kind of verbal language. As shown in the SCM, perception/action and self/other seem to share common informational spaces. Therefore, the actions of others always influence our own behavior, either by priming a similar action in ourselves—that can be inhibited or not—and allowing us to learn new possible actions. Action in this view is not restricted to classical examples, such as throwing balls and grasping objects. This notion can be extended to more complex patterns of behaviors, such as rituals and disciplinary processes.

We are not affirming that language does not play an important role in the cultural exchange; however, we are affirming that nonverbal processes might also have an important contribution that has been underestimated. These body properties are explored by some groups as the main cultural investment. For instance, contemporary anthropological studies state that for Amerindian populations, to be part of the society is to construct a similar and at the same time unique body through social relation with its members (cf. Viveiros de Castro, 2006). Great value is conferred to sharing the same space, participating in social events, memorizing the discourse of the ancients, using substances that affect the perception, paying attention to the subtle variations of the environment, and so on. Gallese (2001) pointed out that there is a neuronal substrate that allows the perception of others' actions as meaningful and addressed to objectives. Therefore, the body cannot be understood as a strictly biological entity, but it is symbolically constructed and disciplined by recursive channelizing of perception and actions.

Exploring the world with its body, the subject gets used with others places through a process of familiarization that entails changes (cf. Boesch, 1991). Thus, if our living is in movement towards the unknown—at least to an unpredictable future—our meaning construction is made as an effort of novelties organization in a previously open-ended configured system (cf. Valsiner, 2001, 2007a, 2007b). This organization is linked to the sensible levels of body excitation and inhibition that entails a feeling tone, from which further semiotic generalizations can develop. Even so, constructed generalized symbolic cultural meanings act recursively on the body, also affecting its felling capacities. Nevertheless, cultural psychology still demands studies on how this recursive process including the body precisely happens. We argue that further studies based on models as Hurley's shared circuit model in association with the understanding of semiotic affective-cognitive constraining processes can be powerful in this direction, overcoming dichotomies such as nature–culture and mind–body.

AKNOWLEDGEMENTS

The preparation of this work was supported by the State of São Paulo Research Foundation fellowship.

NOTES

1. The term *mindreading* is used throughout this chapter as the ability to understand the behavior of others in terms of their mental states, to read intentions in the mind of others (Gallese, 2001)—for example, the ability to tell whether a given observed behavior is the result of an intentional attitude or the consequence of some accidental event. It has by no means any relation to telepathy or other kinds of paranormal phenomena.

REFERENCES

Almeida, A. V., & Falcão, J. T. R. (2008). Piaget e Teorias da Evolução Orgânica. *Psicologia Reflexão e Crítica, 21*(3), 525–532.

Albert, B. (2002). O ouro canibal e a queda do céu: uma crítica xamânica da economia política da natureza (Yanomami). In B. Albert & A. R. Ramos (Eds.), *Pacificando o branco: cosmologias do contato no norte-amazônico* (pp. 239–274). São Paulo, Brazil: Editora UNESP.

Baldwin, J. M. (1896) A new factor in evolution. *American Naturalist, 30,* 441–451, 536–553. Retrieved from http://www.brocku.ca/MeadProject/Baldwin/Baldwin_1896_h.html

Baldwin, J. M. (1894) Imitation: A chapter in the natural history of consciousness. *Mind (new series) 3*(9), 26–55. Retrieved from https://www.brocku.ca/MeadProject/Baldwin/Baldwin_1894.html

Baldwin, J. M. (1906a). *Mental development in the child and the race* (3rd ed.). New York, NY: Macmillan. Retrieved from http://www.brocku.ca/MeadProject/Baldwin/Baldwin_1906/Baldwin_1906_toc.html

Baldwin, J. M. (1906b). *Thought and things: A study of the development and meaning of thought, or genetic logic. Vol. 1, Functional logic, or genetic theory of knowledge.* London, England: Swan Sonnenschein.

Baldwin, J. M. (1930). James Mark Baldwin. In C. Murchison (Ed.), *A history of psychology in autobiography* (Vol. I, pp. 1–30). Worcester, MA: Clark University Press.

Bateson, G. (1972). *Steps to an ecology of mind: Collected essays in anthropology, psychiatry, evolution and epistemology.* New York, NY: Jason Aronson.

Boesch, E. E. (1991). *Symbolic action theory and cultural psychology.* Berlin, Germany.

Boesch, E. E. (2001). Symbolic action theory in cultural psychology. *Culture & Psychology,* 7(4), 479–483.

Chartrand, T., & Bargh, J. (1999). The chameleon effect: The perception–behavior link and social interaction. *Journal of Personality and Social Psychology, 76,* 893–910.

Darwin, C. (2006). *On the origins of species: By means of natural selection.* Mineola, NY: Dover. (Original work published 1859)

Diriwächter, R., & Valsiner, J. (2008). *Striving for the whole: Creating theoretical syntheses.* New Brunswick, NJ: Transaction Publishers.

di Pellegrino, G., Fadiga, L., Fogassi, L., Gallese, V., & Rizzolatti, G. (1992). Understanding motor events: A neurophysiological study. *Experimental Brain Research, 91,* 176–180.

Duran, A. P. (2004) Relação eu—outro: Um paradigma sob a perspectiva do construtivismo terapêutico. In L. M. Simão & A. M. Martínez (Eds.), *O outro no desenvolvimento humano* (pp. 131–144). São Paulo, Brazil: Pioneira Thomson Learning.

Fogel, A., Koeyer, I., Bellagamba, F., & Bell, H. (2002). The dialogical self in the first two years of life: Embarking on a journey of discovery. *Theory and Psychology, 12*(2), 191–205.

Frith, C. (1992). *The cognitive neuropsychology of schizophrenia.* Mahwah, NJ: Erlbaum/London, UK: Taylor & Francis.

Gallese, V., Fadiga, L., Fogassi, L., & Rizzolatti, G. (1996). Action recognition in the premotor cortex, *Brain, 119,* 593–609.

Gallese, V (2001). The 'shared manifold' hypothesis: From mirror neurons to empathy. *Journal of Consciousness Studies, 8*(5–7), 33–50.

Guimarães, D. S. (2011). Amerindian anthropology and cultural psychology: Crossing boundaries and meeting Otherness' worlds. *Culture & Psychology, 12*(2), 139–157.

Hermans, H. J. M, Kempen, H. J. G., & van Loon, R. J. P. (1992). The dialogical self: Beyond individualism and rationalism. *American Psychologist, 47*(1), 23–33.

Hurley, S. (2006). Active perception and perceiving action: The shared circuits hypothesis, perceptual experience. In T. Gendler & J. Hawthorne (Eds.), *Perceptual experience* (pp. 205–259). Oxford University Press.

Hurley, S. (2008). The shared circuits model (SCM): How control, mirroring, and simulation can enable imitation, deliberation, and mindreading, *Behavioral and Brain Sciences, 31,* 1–58

Ingold, T. (2000). *The perception of the environment: Essays on livelihood, dwelling and skill.* New York, NY: Routledge.

Josephs, I. E. (2000). Feeling as movement from a person-centered standpoint: Going beyond William Stern. *Theory & Psychology, 10*(6), 815–829.

Lagrou, E. (2007). *A fluidez da forma: arte, alteridade e agência em uma sociedade amazônica (Kaxinawa, Acre).* Rio de Janeiro, Brazil: Topbooks

Lyons, D. E., Young, A. G., & Keil, F. C. (2007). The hidden structure of overimitation. *Proceedings of the National Academy of Sciences, 104*(50), 19751–19756.

Marková, I. (2000). AmÈdÈe or how to get rid of it: Social representations from a dialogical perspective. *Culture & Psychology, 6*(4), 419–460.

Marková, I. (2006). *Dialogicidade e representações sociais: As dinâmicas da mente.* Petrópolis, Brazil: Vozes.

Meltzoff, A. (2002). Elements of a developmental theory of imitation. In A. Meltzoff & W. Prinz (Eds.), *The imitative mind* (pp. 19–41). Cambridge, UK: Cambridge University Press.

Oppenheimer, L. (1991). The concept of action: A historical perspective. In L. Oppenheimer & J. Valsiner (Eds.), *The origins of action: interdisciplinary and international perspectives* (pp. 1–35). New York, NY: Springer-Verlag.

Rizzolatti, G., Fadiga, L., Gallese, V., & Fogassi, L. (1996). Premotor cortex and the recognition of motor actions. *Cognitive Brain Research, 3,* 131–141.

Salvatore, S., & Valsiner, J. (2009), Idiographic science on its way: Towards making sense of psychology. In S. Salvatore, J. Valsiner, S. Strout, & J. Clegg (Eds.), *YIS: Yearbook of idiographic science* (Vol. 1, pp. 9–19). Rome, Italy: Firera

Seeger, A., Da Matta, R., & Viveiros de Castro, E. B. (1979). A construção da pessoa nas sociedades indígenas brasileiras. *Boletim do Museu Nacional, 32,* 2–19

Simão, L. M. (2003). Beside rupture—Disquiet, beyond the other—Alterity [Versão eletrônica]. *Culture & Psychology, 9*(4), 449–459.

Simão, L. M. (2005). Bildung, culture and self: A possible dialogue with Gadamer, Boesch and Valsiner? *Theory & Psychology, 15*(4), 549–574.

Simão, L. M. (2010) *Ensaios Dialógicos; compartilhamento e diferença nas relações eu outro.* São Paulo: HUCITEC.

Taylor, A-C. (1984). The soul's body and its states: An Amazonian perspective on the nature of being human. *Journal of the Royal Anthropological Institute* (N.S.), *2*(2), 201–215.

Valsiner, J. (1998). *The guided mind: A sociogenetic approach to personality.* Cambridge, MA: Harvard University Press.

Valsiner, J. (2000). *The social mind: Construction of the idea.* Cambridge, UK: Cambridge University press.

Valsiner, J. (2001). *Comparative study of human cultural development.* Madrid, Spain: Fundación Infancia y Aprendizaje.

Valsiner, J. (2007a). *Culture in minds and societies: Foundations of cultural psychology.* Thousand Oaks, CA: Sage Publications.

Valsiner, J. (2007b). Human development as migration: Striving toward the unknown. In L. M. Simão & J. Valsiner (Eds.), *Otherness in question: Labyrinths of the self* (pp. 349–378). Charlotte, NC: Information Age.

Valsiner, J. (2008). Baldwin's quest: A universal logic of development. In J. Clegg (Ed.), *The observation of human systems: Lessons from the history of anti-reductionist empirical psychology* (pp. 45–82). New Brunswick, NY: Transaction Publishers.

Viveiros de Castro, E. B. (2002). *A inconstância da alma selvagem e outros ensaios de antropologia.* São Paulo, Brazil: Cosac Naify.

Vygotsky, L. S., & Luria, A. R. (1996). *Estudos sobre a história do comportamento: o macaco, o primitivo e a criança.* Porto Alegre, RS: Artes Médicas (Original published in 1930).

CHAPTER 6

EXPLORING EXPLORATION AS A RECURSIVE PROCESS

Meike Watzlawik
Clark University, United States

Elli Schachter
Bar Ilan University, Israel

Carla Cunha
University of Minho & ISMAI, Portugal

ABSTRACT

Exploration is one of the core concepts in identity research. Identity theorists thus far mainly concentrate on exploration as the process in which future selves are compared to current selves. This process is triggered in a number of ways. In this chapter, we will emphasize culturally normative as well as non-normative identity exploration among adolescents and young adults. Through analysis of case studies, we find that people not only compare future to current, but also past selves to current selves, and in the process revise their identity histories. We explore how recursivity and revisiting inform the exploration process and do this by looking not just into the future, but backward, too, which is a process we know relatively less about in identity literature.

Cultural Psychology of Recursive Processes, pages 163–194

WHAT EXACTLY IS RECURSIVITY?

In this chapter, we want to examine how recursivity can help us understand identity development. Before defining what we mean by "identity," itself being a term with many definitions, let us thus first ask: What exactly is recursivity? The question is not easy to answer. We can start with what it is not: It is not "simple" repetition. Repeating something means to do the same again: A person can repeat push-ups during a workout, the TV program is repeated the day after, or the researcher repeats an experiment. Especially for the last example, it seems essential that—if the stability of the results is to be tested—the experiment is conducted in exactly the same manner as it was the first time (replication).

Is that possible? A chemist who tests how two different substances react under certain conditions can certainly make sure that, when trying to replicate the findings, he uses exactly the same substance amount, the room has exactly the same temperature, and the utensils are sterile as they were the first time, and so on. What has changed for sure, though, is the researcher himself. By knowing what the results of the first experiment were, the meaning-making processes (within the person, in this case the chemist) that accompany the repeated realization of the experiment have altered. The researcher's goals have changed. Driesch (1925) even states that meaning making processes are always dynamic, they never repeat themselves, and thus, meaning does not repeat itself. In his book *The Crisis in Psychology*, he argues, for example, that we do not "have" hope. With every situation in which we hope for something, the hope is different. We do not have the same hope again. Instead, we experience always another "something." Driesch (1925, p. 25) explains this statement as follows:

> I never can have the very same content a second or third time, because, by its having been had already, it is made different from what it was the first time! For the second or any subsequent time, that content carries in itself two accents: one of before and another of already known, which it did not carry when it was possessed first. Thus every content is exclusively what it is and there cannot be two quite identical contents.

In principle, "simple" repetition is consequently not possible, when we consider the meaning making processes that accompany these experiences. A person can show the same behavior (A), but the way this person (I) creates meaning (P), while repeatedly showing A, is not the same.[1] This point being made is not a new one. Heraclitus (c. 535–c. 475 BCE) has become famous for his sentence: "No man ever steps in the same river twice, for it's not the same river and he's not the same man." This is even true for simple things such as the repeated push-ups during work out: The person might be fresh and motivated at the first set but has to increase the effort with

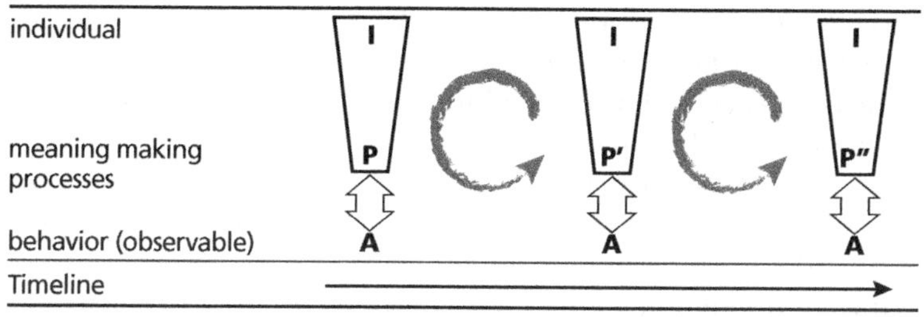

Figure 6.1 Meaning-making processes with recourse to past events while the observable behavior stays the same.

every push-up.[2] At a closer look, we therefore see that individuals always connect the present with the past: They create meaning with *recourse* to past events/experiences. Clifford, Friesen, and Jardine (2001, p. 2) even refer to recursivity in the humanities as "the act of a mind or self 'looping back,' 'turning around' or reflecting on itself, and in this way actually creating itself as a conscious self—the highest expression of human awareness." This, of course, takes place within a certain social system providing feedback to the individual that is reflecting on itself as part of this system.

When dealing with humans, we automatically deal with recursive processes (see Figure 6.1)—even if they do not lead to behavioral changes.[3]

IDENTITY = IDENTITY?

Another, more difficult question is, if the person in Figure 6.1 really stays "the same" over time? If every aspect of the person is to remain the same in order for the person to be considered the same, then we are—taking Driesch and meaning-making processes into account—constantly changing and so identity as "objective" sameness is impossible.

Important to note for identity research is that not only is the object changing, but more important for identity research is that the subject *looking at the object* is changing as well. Therefore, identity in the psychological sense refers not to objective sameness but the attempt to impose and construct a semblance of stability on something dynamic. Change needs to be dealt with, and identity is a way of giving stability to the flux. Therefore, one way of looking at identity is that it is a construction trying to "freeze" aspects of change, or find threads of continuity in change, so as to allow meaning.

The social context or societies individuals belong to provide (and enforce) many frameworks as resources for creating stability out of flux: National identity (e.g., Smith, 1993), social identity (e.g., Jenkins, 2008), cultural identity (e.g., Hall & du Gay, 1996), professional identity (e.g., Yang

Costello, 2005), gender identity (e.g., Maccoby, 1999), sexual identity (e.g., Watzlawik, 2004a), or even place identity (e.g., Sarbin, 1983) are a few of the constructs with which an individual (or community) can position him- or herself (itself). What all of these constructs have in common is that they describe identity structures,[4] or to be more precise: The constructs are, if needed, used to provide individuals (or researchers) with a sense of structure and meaning within social frameworks.

Identity development is thus the continuous effort to make sense of change by creating tentative understandings of reality and reifying these. However, since reality does not stand still, the object *and the understandings* constantly need to be revisited demonstrating the recursivity of the process.

EGO-STRUCTURE: PROCESS VARIABLES AND IDENTITY STATUS

Marcia (1993) assumed that identity serves to *structure* an individual's psychodynamic processes (e.g., aspirations, skills, beliefs, individual experiences); thus, he defines identity as ego-structure: an internal, self-constructed, dynamic organisation. Observable *behavior,* he assumes, is a reflection of how an individual is trying to make sense of her-/himself and his/her environment.[5] The *experience* of one's own identity, however, includes the *process* with which individuals become aware of certain traits/characteristics (e.g., being part of a certain family, having certain abilities, falling in love with persons of the same or opposite sex). If the individual has no active part in this process, Marcia (1993) talks about identity *formation*; if the individual plays an active part, identity *construction* is taking place. Both can either be triggered by biological maturation or/and the insertion into cultural and social institutions (cp. Erikson, 1968).

Following Erikson, Marcia (1980) considers the life phase adolescence[6] as crucial for identity development—even though identity development is a lifelong endeavor (Stephen, Fraser, & Marcia, 1992). He assumes that adolescents experience an identity crisis, in which former elements of the ego-structure are questioned and reconstructed, and new aspects are or have to be discovered, integrated, or configured (cf. Schachter, 2002; Watzlawik, 2004b). Crises are resolved by making choices (commitment[7]) in a variety of life domains. Marcia (1966) uses these two core concepts—crisis and commitment—to define different identity development statuses (see Table 6.1).

The term *crisis* implies that (1) choices have to be made, and that (2) the individual is in a situation in which he is "seriously psychological disturbed" because he has difficulties choosing (Meeus, 1996, p. 97). However, because the second could not be proven empirically (Offer, 1969, as cited in Meeus, 1996[8]), many authors (Grotevant & Cooper, 1986; Meeus, 1996), including

TABLE 6.1 Marcia's Identity Status Model (Marcia, 1966)

	Statuses			
Variables	Identity Diffusion	Foreclosure	Moratorium	Identity Achievement
Crisis	Yes or No	No	Actual	Yes, past
Commitment	No	Yes	Unclear	Yes

Marcia himself (see, for example, Marcia, 1987), replaced (specified) the term *crisis* with "exploration." Exploration thus stands for the evaluation of a variety of alternatives in different life domains (e.g., politics, religion, profession), or, as Grotevant (1987, p. 204) states: It is "'problem-solving behavior aimed at eliciting information about oneself or one's environment in order to make a decision about an important life choice."

Marcia (1987) considers exploration and commitment to be *process variables.* Identity status is then determined by the degree of exploration and the subsequent commitment (Marcia, 1987); both can, for example, be assessed with the Identity Status Interview (ISI; Marcia, 2007), which includes questions like "How did you come to decide upon [field of study/profession]?" or "What other things have you considered besides [field of study/profession]?" Although these interviews lead to quite detailed process descriptions, the data is, nevertheless, (only) used to determine states/statuses (e.g., X is currently in the identity status Moratorium)[9] in different domains. Since the identity statuses can vary in different domains, we should probably not talk about identity but identities—an important point being made, because identities can contradict each other, but we will go into this later. Since we are mainly interested in the process and not identity status, identity structure, or changes in either (see, for example, Moratorium-Achievement [MAMA] cycles, Stephen, Fraser, & Marcia, 1992), we will have a closer look at the process variable exploration.

EXPLORATION

Flum and Kaplan (2006) try to summarize what different theoretical concepts of exploration have in common. They say that exploration is about "engagement with the environment and the motivation to acquire information [in relation to the self] through interaction with the world" (p. 100). Exploration is thus a process of examination and discovery of who and what an individual might be (Berman, Schwartz, Kurtines, & Berman, 2001). Exploration is commonly conceived as an innate motivational propensity (Bowlby, 1969) and hence as primarily intrinsically motivated (Deci &

Ryan, 2000, as cited in Flum & Kaplan, 2006). Different types of exploration can be distinguished if we consider their purpose: Whereas Marcia defines exploration mainly as evaluating various alternatives before a choice is being made (Marcia, 1987, 1993), Meeus, Iedema, and Maassen (2002) define exploration as the extent to which adolescents (and adults) currently think and actively gather information about their commitments (choices made). Thus Marcia focuses on the formation (construction) of commitments, whereas Meeus et al. concentrate on the maintenance of commitments through exploration (Luyckx, Goossens, Soenens, & Beyers, 2006). With these different foci, it seems reasonable to differentiate between different kinds of exploration (Grotevant, 1987; Marcia & Archer, 1993): exploration in depth and exploration in breadth. The latter stands for the gathering of information about different identity alternatives; exploration in depth captures the gathering information about current choices (cp. Luyckx et al., 2006). For validation purposes, both kinds of exploration were assessed with the help of questionnaires (Luyckx et al., 2006). For exploration in depth, research participants evaluated statements like "I think a lot about my education" or "I try to figure out regularly what other people think about my best friend"; for exploration in breadth, sample items were "I have considered adopting different kinds of religious beliefs" or "I have evaluated many ways in which I fit into my family structure." The results that are based on these quantitative assessments confirmed the assumed differentiation of exploration. However, Luyckx et al. (2008) believe that both types do not capture exploration sufficiently. They suggest adding a third type that they call ruminative or maladaptive exploration.[10] In this case, the individual is "stuck" in a cyclic process (see *feedback loops* in the next section) characterized by a repetitive and passive focus that makes him or her feel hopeless and out of control of the situation at hand.

All three types are future- and present-oriented. An exploring individual is portrayed as gathering information for a future decision in relation to the self—yet there is not much discussion of what is the way the self is looked at in this process—and not only the external alternatives. We examine this by looking at other important questions in debate.

What Triggers Exploration (and Keeps It Going)?

First of all: Why would somebody *start* exploring? As mentioned above, identity issues can, for example, become relevant due to biological maturation or/and the insertion into cultural and social institutions. In adolescence this may lead to questions about "whom to date, whether or not to break up, having intercourse, taking drugs, going to college or working, which college, what major, studying or playing, being politically active, and so on" (Marcia, 1980, p. 161)—all having implications for who adolescents are and want to become. Even beyond adolescence, biological changes can also trigger identity issues, such as menopause, aging (needing glasses, getting wrinkles, etc.), and cultural and societal norms can trigger identity formation/construction (e.g., retirement regulations, expectations about appropriate behavior).

However, there is another approach that goes beyond biological changes and the insertion into cultural and societal norms—not without including both. Authors have pointed out that disruptions (Bosma & Kunnen, 2001) or rupture-transitions (Zittoun, 2006) can trigger exploration. Ruptures are interruptions of the normal flow of events, which can happen throughout the life course. These interruptions are very diverse in nature. They can, for example, be caused by perceptions that contradict current self-perceptions or by life events ("personal catastrophes") that alter the current system (e.g., death of a loved one, illness) so that it must be modified until relative stability (congruence) is achieved (disequilibration-equilibration-process, cp. Zittoun, 2006). Coming back to the above-mentioned "identities" a person can have and that Marcia defined by examining different life domains, contradictions are, of course, part of an individual's everyday self-experience. A man can describe himself as family man and can still be a workaholic; a professor can be committed to his students and be committed to his research at the same time, self-descriptions and experiences change with the context, but this does not prevent the individual from having an overall feeling of sameness or congruence—along with the task of juggling these identities (e.g., "How can I credibly state that I am a family man when I am again spending my night at the office?"). Individuals prioritize no question. However, Zittoun is talking about a different kind of contradiction. What happens, for example, if the above-mentioned workaholic and family man starts having an affair with another woman, risking the trust of his wife (family)? Or more tragic: What if he loses his family in a car accident? These contradictions or life-events will then lead to modifications in the current system until the system is (hopefully) stabilized again. Transitions are thus the process of restructuring. This process includes meaning-making. According to Zittoun (2006), meaning is established along two axes: (1) meaning-making by the linking of time and (2) meaning-making by relating events (ruptures) to a system of orientation that includes values and criteria on which thinking and actions are based (the "piloting system," Zittoun, 2006, p. 191). The linkage in time is achieved by maintaining a sense of the past and by presently working towards the future. In this process the representation of the past underlies changes. Josselson (2009) has elaborated on this further. With the help of case studies, she demonstrates how the present in fact constructs the past and shows "how autobiographical memory may be used dialogically to create and contrast with current self-constructions, to disavow intolerable aspects of self, and to preserve disused but valued self-representations" (p. 647).

Meaning-making processes also include emotional regulations and elaborations of unconscious processes awaken by the rupture (Perret-Clermont & Zittoun, 2002, as cited in Zittoun, 2006). Exploration is part of the meaning-making process. Individuals have to figure out what happened, what the consequences are, how they feel, and what it means for who they are or want to be in relation to who they once were; they have to discern what their alternatives are—and this is a rather complex endeavor. By dividing the exploration process into five interacting elements, Grotevant (1987) tried to capture this complexity. He defined exploration as problem-solving behavior and differentiated the following processes/factors:[11] (1) expectations and beliefs, (2) information gathering and hypothesis testing (actual exploration), (3) investment of time and effort, (4) forces that compete with the exploration process for the person's attention and effort, and (5) provisional evaluation. Feedback loops (which can be seen as recursive processes) between (2) and (5) characterize the exploration process. "Depending on whether there is a good of fit [sic] between the new sense of identity and the environment, and on the satisfaction given by that new identity, exploration is either continued or not" (Bosma & Kunnen, 2001, p. 53).

Grotevant's model (1987) still leaves questions unanswered (Kerpelman, Pittman, & Lamke, 1997): What causes the exploration process to continue? What occurs during consolidation? With these questions, Kerpelman et al. (1997) stress the need to analyze the microprocesses that underlie identity exploration (and commitment), and they do so in their identity control theory (ICT) approach. The approach is based on self-regulating interpersonal (social feedback) and intrapersonal components (self-definitions). A negative feedback loop minimizes the discrepancy between the two. The loop can also be seen as attempt to verify one's own identity. The authors describe this loop as follows:

> [I]nterpersonal feedback is received (A) and interpreted to become a self-perception that is matched (B) by a comparator with input (C) from an identity standard. When the standard and self-perception are incongruent, an error/disturbance results that leads (D) to the enactment of behavior aimed at restoring the predisrupted identity. For this restoration to take place, cognitive behavior may shape self-perceptions directly (E), or social behavior may change the interpersonal situation (F) leading to new social feedback (A). The original identity standard is maintained when behavior produces congruence between self-perception and identity standard. However, when behavior fails repeatedly to result in congruence between self-perception and the identity standard, an alternate means of reestablishing congruence is to adjust the identity standard itself (G). (Kerpelman et al., 1997, p. 329)

The individual constantly monitors congruence between input and standard, but exploration (as part of the control apparatus) is only activated in case of a discrepancy. Kerpelman (2001, p. 83) explicitly defines exploration as "those thoughts and behaviors in which a person engages as he or she tries to obtain [social] feedback that he or she will interpret to be consistent with who he or she intends to be." Importantly, social others as well as cognitive processes within the individual—trying to behave consistently with anticipated identities—can be the source for this feedback.

There are multiple factors that influence the activation of the control apparatus and, thus, exploration and identity development: The amount of discrepancy, the importance of and the confidence in the identity standard (see next section), the type of feedback (accurate, flattering), social context of the feedback (a congruent or an incongruent partner) (Bosma & Kunnen, 2001), the capacity to withstand guilt and fear (Kroger, 2006), the style of reasoning (Stephen et al., 1992), the information processing style (Berzonsky, 1989), and problem-solving competence (Berman et al., 2001) are only some of them. Once again, however, the ICT model has the exploring individual compare the feedback with what a person wants to be, or believes she is.

However, ICT does not address the issue of what the person believes she is and what processes have been experienced so as to change that perception while still retaining a sense of continuity.

In the following section, we present this alternative portrayal of exploration as a continuous ongoing process that also involves a revisiting of past identity constructions.

How New Construction Builds on Revisiting Past Ones

We said that the subjective sense of identity can be seen as structure. We then have described some of the processes that have been offered to explain how this structure develops, focusing on exploration and thus periods of change. The term *structure* implies that there are several components that exist, creating a sense of congruence.

When examining identity structures or developmental processes concerning identity, researchers usually concentrate on specific life domains or identity aspects, and we do so here, too, taking the example of gender identity to demonstrate our point. The numeric aspect is already determined before we are born. The genetic code determines whether we are categorized as male or female or something in between.[12] After being born, we are treated a certain way and are confronted with expectations concerning the different sexes, which also shape our gender identities. During this time, we constantly monitor the congruence between identity standards and feedback—sometimes more, sometimes less consciously. Gender identity is usually verified in a recursive assessment, in which past and present consistently cohere. Puberty, however, can be seen as a rupture-transition that forces us to deal with physical issues (e.g., accepting bodily changes). To maintain stable identity, we have to integrate these changes into our *former* gender identity and modify it in this process. With this example, it becomes obvious that the future-oriented approach on exploration in identity development falls short. Only with this looping back does identity development become understandable. This looping back can also be done to establish *meaning bridges* (Brinegar, Salvi, Stiles, & Greenberg, 2006). A meaning bridge expresses an understanding or reconciliation between opposites (e.g., contrasting perspectives between self and other or between parts of oneself, as one discovers discrepancies or incongruence in identity). Several authors consider it a powerful semiotic tool in human development, since this processes involve meaning-making efforts that provide a connection between former and present identity structures, allowing to achieve new forms of self-integration and reconciliation (Brinegar et al., 2006; Cunha et al., 2013).

After consolidation, unexpected events can again cause incongruence (e.g., pregnancy, discovering homosexual feelings, a man is found to be infertile, menopause) and the former identity construction must be re-examined. The same chain of stability and change is possible in other aspects of identity—either simultaneously or time-delayed. Interaction between the different aspects is also possible; for example, has the discovery of being homosexual (part of one's sexual identity) also influenced one's gender identity (e.g., "Am I still a real man when I am sexually attracted to men?")? Other aspects might be left unaffected. These examples show that we constantly "work" on our identities, establishing meaning bridges, backwards and forwards in time and concurrently between different domains. The different aspects exist in parallel and can be stable or in transition. They can also be central for our identities or only of marginal importance, which also affects the intensity with which we invest time and effort. Figuratively speaking, if different interacting identity aspects/facets develop in parallel, then identity structure would be the cross section at a specific point in time. Identity development (processes) can, in contrast, only be captured longitudinally.

CASE STUDIES

We have shown that exploration is far more complex than walking into the world and evaluating future alternatives in relation to the present self. Individuals experience incongruence and can actively search for alternatives—but also need to recursively re-evaluate previous constructions in order to maintain a sense of continuity. Recursion within or, better, as part of identity development is characterized, thus, not only as a forward-oriented endeavor, but also a process that includes looking back: The individual experiences something and evaluates this experience by (1) looping-back as the primary means of self-awareness (self-referencing); then, if the experience is not repelled or ignored, (2) change can be observed. As a third step, the new experience is (3) integrated and becomes meaningful for identity. Present (Who am I?), future (Who will I become?), and past (Who have I been?) thus mutually modify each other in this process. Hence, self-aware individuals need to look backward while moving forward. In this sense, recursivity is doubly appropriate: Individuals loop back, as a matter of process, and individuals look back, as a matter of exploration.

The following examples with which we will try to illustrate the above are all taken from different studies on identity development. An overview of the studies' goals, methods, and samples can be taken from Table 6.2. Since all studies focused on personal transitions, recursive processes should become visible in each of them. We will later discuss which implications can be drawn from the findings for our understanding of the process of exploration.

Before describing the first case study, one important point needs to be made. Researchers choose different paths for gathering data. In longitudinal

TABLE 6.2 Overview of Studies from Which the Following Examples Were Taken

	Israel	Germany	Canada
Who was interviewed?	Miri	Homosexual and bisexual adolescents	Sarah
Study's goal	Life-story research, investigate identity development	Investigate sexual identity development in adolescence	Investigate narrative change processes in Emotion-Focused Therapy for depression
Reference	Schachter (2004)	Watzlawik (2004a)	Cunha et al. (2013)
Total sample size of study (N)	30	1283	6
Data analyzed	interview transcriptions	written answers to an online questionnaire	transcriptions from therapeutic sessions (dialogues with emphasis on the exploration and transformation of meanings clients attribute to their life problems)

studies, development in assessed "while it happens," so that change can be monitored. In cross-sectional studies, participants are usually asked to assess either a current state or describe an event retrospectively. In all three studies presented here, the participants describe past events that triggered change. If our argumentation is correct, our interviewees should—when exploring the options—not only mention effects this event had on them and the decisions that *followed*, but they should also refer to feelings/events/constructs that they had or applied, or that occurred *before* the event (looping back to make sense out of the *then* current situation). The "problem" with this approach is that memories or interpretations of past events are certainly influenced by experiences the individual gathered thereafter (looping back as a constant source for information). In all studies, we thus focus on how our research participants describe these interpretational processes today as part of the meaning making involved.

MIRI: THE EXPERIENCE OF IDENTITY RUPTURES (ISRAEL)

The following excerpt is part of a life-story research interview. In the interview, the interviewee, Miri, a 24-year-old woman, tells a story that leads up to two specific episodes that take place when she is around 20 years old. The stories, which we will soon relate, are of identity exploration triggers and the ensuing exploration.

One way to relate to a life-story is as a straightforward depiction of past events. If we do so, then, in our case, this excerpt could be a way to get a firsthand account of Miri's actual past exploration. Miri is reporting now about event A, her feelings at the time, and the following identity crises and reflections. It is a story of how she explored event A at point B somewhat afterwards. We will take a different approach, though, and look at the story as a narrative framed within the interaction between interviewee and researcher (Bamberg, 2010). As such, we can understand her story as an act of interpretation of past events that is taking place now with the interviewer. What we then have before us is Miri exploring at point C her exploration of event A at point B. We will present this as an example of a recursive process. In the first loop in point B she looks at A. Later on, at point C, she is looking at her previous construction.

Miri describes an identity story that takes the classic form described before by Marcia: The experience of wholeness is interrupted by a crisis and exploration follows. What is interesting is her implicit awareness that she is actually now fitting the particulars of the story into this template, as we shall soon see. She begins the story describing herself as a religious teenager—very serious about her religiosity. Her religion is described as something very "whole" and "complete." However, she then describes a feeling of a loss of vitality in her religious identity. We highlight the words we especially wish to analyze in italics.

> *What's more important for the continuation of the story* is that, at this period [childhood and adolescence], my world was very religiously complete. I loved it, too. I felt it gave me a comprehensive picture on life—*well I don't know to what extent I felt this at every single age*—but at least I had a will to be part of this [religious] world which stemmed from my feeling that this world is true and meaningful and enables one to develop and advance. [Emphasis in this and subsequent passages has been added]

Notice the depiction of her feeling then that religion gave her a comprehensive picture on life—and then a slight backtrack. Miri tries to be more precise and qualifies her previous sentence by saying that she is not sure whether this feeling was there to the same extent at every age. Miri is showing awareness that there is a discrepancy between what might really have been (i.e., some variability) and the aggregate that she wants to present now as a constant. She expressly understands that this aggregation is for the purposes of the story she is telling. She says in the opening sentence "What's more important for the continuation of the story is that..." showing that the ensuing aggregate is done purposefully so as to create a constancy that will make sense of her later feeling of disruption. This shows that she is constructing the story using a template of consistency followed by rupture—because that template makes sense as a story; however, she also wants to be loyal to the "truth" as it "really happened."

This awareness continues further on in the story when she reaches the episode of the crisis. Miri describes how she joined the army after high school, although religious women in Israel are granted an exemption from conscription if they so request, as this is considered by some not religiously proper. She became an officer and served as an instructor. She saw herself as a (unofficial) representative of the religious community in the army, and as a sort of "spokeswoman" in the eyes of nonreligious soldiers. However, in the course of her interview she recounts a few stories of a realization that this is changing:

> And what *gradually happened—one day* towards the end of my service something interesting happened. I was sitting with a religious friend of mine, we were working on some project, and there were two other officers who were designing a wall poster [to hang up on the barracks wall] about the Passover holiday. And they turned to us and asked: "By the way, why is Passover also called the 'Holiday of Freedom'?" And we smiled at each other and we chanted the certified answer from our schooldays: "*Freedom? We left [Egypt] to the only real freedom there is, [which is in] keeping G-d's commandments*"[13] And we said it in such a cynical tone. *And I realized—we both realized— that something happened to us in the army*, that *all of a sudden* we are talking about these things

> in dreadful cynicism, *that all of a sudden* all this school poppycock didn't seem convincing. *And that was a great shock to me to feel that I wasn't the same person on the inside that I used to be....* You see, the world *I am in now* strengthens my critical faculties and not—I'm not connected from the inside to the religious world. I don't feel—do you understand what I'm saying? [Interviewer: What *do* you mean by that?] My religious soul wasn't there anymore, my religious happiness, my innocent faith, something like that.... And this feeling made me want to go study again, to re-examine myself, to explore my faith, to bring my worldview up-to-date.

Here Miri's description of this episode begins with the terms *gradual* and *sudden.* The underlying change is gradual, and yet she describes the *realization* of change as sudden. This is a retroactive interpretation—the sudden realization of identity change causes the interpretation of the change being gradual—for only if the change was gradual could it have happened unnoticed and so surprise her. If there was no surprise, then there would be no realization of gradual change. Yet from here, Miri continues to speak of the world she is in "now" strengthening her critical faculties at the expense of an "inside connection" to the religious world. Here, we have another third "current" vantage point. Now, she looks back at her interpretation and deepens it attributing the gradual change to an increase in critical faculties. Later on she says:

> What happened to me afterwards, religiously, happened—well there are all sorts of levels that are dependent one on the other. After the incident about Passover. This made me realize I want to study again. *To go back and re-examine myself.* [Interviewer: Re-examine yourself or...?] Myself. Myself, my faith, my worldview. To create a worldview—more up-to-date than the one I had. I understood that I hadn't moved forward since my schooldays—not spiritually at least. If the answers I had were good for school and I don't believe in them so I need to build something new or strengthen what I've got or check out what is there, at least some sort of renewal.

In this passage, Miri looks back and attributes her later embarking on a moratorium period of study to the realization made then—yet once again, as narrator now, she frames this as being more complex—as being part of "all sorts of levels that are dependent one on the other." From her current vantage point, she is recursively going back to her previous meaning-making adding/subtracting meanings. What is important to point out is that she does not go back to the raw experience, but to the meanings already formed then. They get looked at once again and are reorganized.

Later in the interview she introduces the next story by saying: "But what *really, I think,* was the turning point in this regard was.... " Notice how she attempts to give the story a stamp of veracity—"really"—coupled with a qualification "I think." Retroactively she knows she has changed gradually, yet she is attempting to pinpoint "the real" changing point by reinterpreting past events.

She describes becoming very attracted to a "very" nonreligious attractive colleague. They are "so different" because of their backgrounds and different codes of behavior. After three weeks during which she abstained from even touching him (premarital sex is forbidden, and many Jewish orthodox youth abstain from even touching), she cannot conceive of the relationship having a future and she decides to break it off.

> And then I gave him my hand. After it all [ended] [laughter]. And that [touch] suddenly turned my whole world upside down. Holding hands. *It was a tremendous revelation of a whole new world. It was a turning point.* After holding his hand, after it all, *I suddenly felt that my whole world became unstable, well maybe not exactly at that moment, but afterwards. But that was the moment that changed my whole perspective on this issue.* [Interviewer: What did you realize?] *All of the sudden I understood* that I had mistreated myself beforehand, that this had made me unhappy—not touching—repressing myself so harshly. On the other hand, this undermined my whole religious world because I said to myself if, if it seems so trivial to give my hand to someone to touch, if I see it as so self-evident. *I suddenly understood* that it is necessary. That it is elementary in a relationship between two people. So maybe something is wrong with the religious world if it forbids it so. *Suddenly* the whole fabric of things that had seemed so perfect—got all unravelled. *And another thing that I decided then* was another axiom or basic principal and that was that I am OK about this. That it is very healthy and right to touch someone that I have a relationship with. So what I have to do is to find out is whether the religious world is OK or not or where this fits in the religious world.

The story Miri tells—of a touch that changes her world—is interesting for many reasons. Here we especially point out how the feeling of "perfection" which she previously attributed to herself in the period of childhood and adolescence (which she said that she "loved") turns into "I had mistreated myself," "repressing myself so harshly." She now says that she was then "unhappy." Even while describing the revelation that caused the change in perspective, she does so again in a dual fashion: "I *suddenly felt* that my whole world became unstable, *well maybe not exactly at that moment, but afterwards.*" She attempts to describe the change as a sudden event happening then, but acknowledges that perhaps the realization came afterwards.

Once again, we have an awareness of fitting the complex story into a straightforward narrative form of the classic identity crises. Miri is aware that it might not have been so straightforward—she struggles to explain that it fits in the narrative form she wishes to present as a psychological truth with another view of the facts. It stands to reason that Miri is struggling both with her awareness that certain thoughts might not have gone through her head in the exact way she describes at the exact moments she describes and therefore her account is violating standards of truthful storytelling (or that perhaps they did go through her head but so did other contrasting thoughts)—yet also with the notion that the story tells some sort of psychological truth when looked at from her current vantage point. Possibly she believes that those thoughts were there, but in a less developed form or in a more subdued form. Miri is allowing herself to retroactively flesh out and interpret these undeveloped thoughts as they "really" were. The reality of those thoughts being there is proved retrospectively by the long-term developmental story they effected. If they were not there, then in some basic form she could not explain her later change, and on another level we might say, if they were not there in some basic form, the moment would not have been memory encoded and could not be interpreted later.

If we now explicitly focus our discussion on exploration, what we have here is an identity exploration that is being described and recognized retroactively in a further act of exploration. We might try to ask the question of when the exploration happened. It is problematic to say that it happened when she was 20, for it was then not fleshed out. It is also problematic to say that it did not happen then but happened at the time the story is being told to the interviewer (after what it caused to happen had already happened!), because obviously some meaning making in regards to the self did happen back then—if only fleetingly. A better explanation is to see the exploration process as constantly going on as an interpretive process that at certain times must reinterpret previous interpretations.

SEXUAL IDENTITY: COMING OUT (GERMANY/USA)

Sexual identity as one facet of an individual's identity structure also includes several components: becoming aware of one's sexual attraction toward persons of the same, other sex, or both sexes (mainly during puberty, but not solely), experimenting with intimacy, and finding out more about one's own sexual and emotional needs, as well as taking responsibility for oneself and others. Many adolescents, when first feeling sexually attracted to another person, are confused at first (Watzlawik, 2004b). Same-sex attracted adolescents face additional obstacles. These adolescents have to figure out for themselves what being homosexual or bisexual entails with little help or understanding, and they are often confronted with prejudice and discrimination (Schupp, 1999; Watzlawik & Heine, 2009). The process of becoming aware of one's own feelings includes recursive processes/exploration.

The data presented here were collected in 2000 with the help of an online questionnaire. Among other aspects, the participating German- ($N = 809$) and English-speaking ($N = 474$) adolescents were asked to describe how they discovered their sexual orientations. Their ages ranges from 12 to 16 years. We will examine their answers to depict what kind of recursive processes occur.

Looping Back to Former Events

One of the adolescents described the moment in which he became aware of his sexual orientation as follows:

> Hm ... hard to describe. I was hit with a sudden rush of emotions and memories. I recalled emotions from a scene in the past involving a girl I had a deep crush on. I suddenly realized I had the same feelings, perhaps even stronger for another boy. I thought "Wow, hey, I'm bi." There were a rush of other thoughts, but that was the most clear part. (ID 16)

Only by connecting his current feelings for a boy to those for a girl in the past, he realizes that (a) he has a crush on the boy, and (b) he is bisexual. He does not depreciate former or current feelings, but integrates both into the identity facet "I am bisexual." Another girl says explicitly that past experiences made her realize what was going on: "Well, when I realized what it was and then looked back on the past and the way I felt about some girls (they were pretty but it was 'different') it all clicked" (ID 639).

Whereas in these cases, the process seems to be a rather uncomplicated, others struggle more to accept what they experience:

> I was shocked. Being a devout Christian I was taught that homosexuality was bad (luckily for me, that part I couldn't believe that much, because I know another gay man, and he can't be bad, so . . .). I was wondering what happened, and how I was like that, and if this was just a stage, because I have heard that people are often curious at my age, and usually is just a stage, so I'm not completely sure yet, because I don't know how long stages like this usually last, but it is one long stage. (ID 362)

When this girl realized that she had feelings for (only) other girls, she looped back to what she was taught about homosexuality in church. Since this is a rather threatening thought (she could not accept herself when accepting the religious view on homosexuality), she finds a way to question it (also experience-based). Knowing a homosexual man, who apparently does not fit the picture church draws, helps her to maintain a positive self-perception. Nevertheless, she does not take on the identity facet "I am homosexual" (yet), because she knows that homosexual feelings can be a "stage." If the feelings disappeared, she would not have to struggle anymore, but reading the last sentence she wrote, it seems she doubts this is possible. At a certain point the recurrent attempt to label it a stage will have to be abandoned. Then the recursive looping back will lead to change and a reorganization of identity structures.

In both of these cases, self-perception (having feelings for persons of the same sex) and identity standard (I am heterosexual) were incongruent. Both of the adolescents were shocked about what they felt, but managed to interpret (cognitive behavior) their feelings by recursively integrating past experiences to re-establish congruence (adjust identity standard to either "I am bi" or "I am homosexual"). The process until congruence is re-established equals exploration (evaluating the options). Some adolescents, nevertheless, deny what they feel to maintain their identity standard ("I am heterosexual").

Blocking Self-Perceptions Until They Cannot Be Blocked Anymore

Kerpelman et al. (1997, p. 329) wrote that "when behavior fails repeatedly to result in congruence between self-perception and the identity standard, an alternate means of re-establishing congruence is to adjust the identity standard itself (G)." During the time in which behavior fails to establish congruence, the individual apparently uses different techniques to ignore those self-perceptions that threaten the current standard. One boy writes, "I was telling myself I wasn't gay, I kept making up excuses why I like other boys, but I finally accepted that I was gay about a year ago" (ID 167).

Another girl writes, "I didn't really think about it much I tried to ignore it as much as I could until it got strong enough in sixth grade I couldn't ignore it anymore" (ID 33).

Whereas the boy seems to rationalize what he feels, the girl suppresses her feelings until they cannot be ignored any longer. These techniques can certainly be described as defense mechanisms (Freud, 1937; Vaillant, 1992). The boy in the next example describes a similar process: "Whoa, that's not right, was my initial reaction, then I tried to block it out; this went on for about a year until I finally fully accepted the fact that I was gay" (ID 93).

This boy feels something, explores what these feelings mean, evaluates them, and—since the evaluation is a negative one—"blocks it out." The girl above (ID 33) describes what can happen if one does this more than once: The feeling—even though being ignored—gets stronger until it cannot be ignored anymore. Figure 6.2 shows this accumulation and the exploration processes connected.

The accumulation in Phase I is only possible when the different events that were incongruent with the current identity standard are recursively connected with each other. When the current identity standard is "I am heterosexual," then events that contradict this standard could be falling in love with a person of the same sex, finding persons of the same sex sexually attractive, dreaming

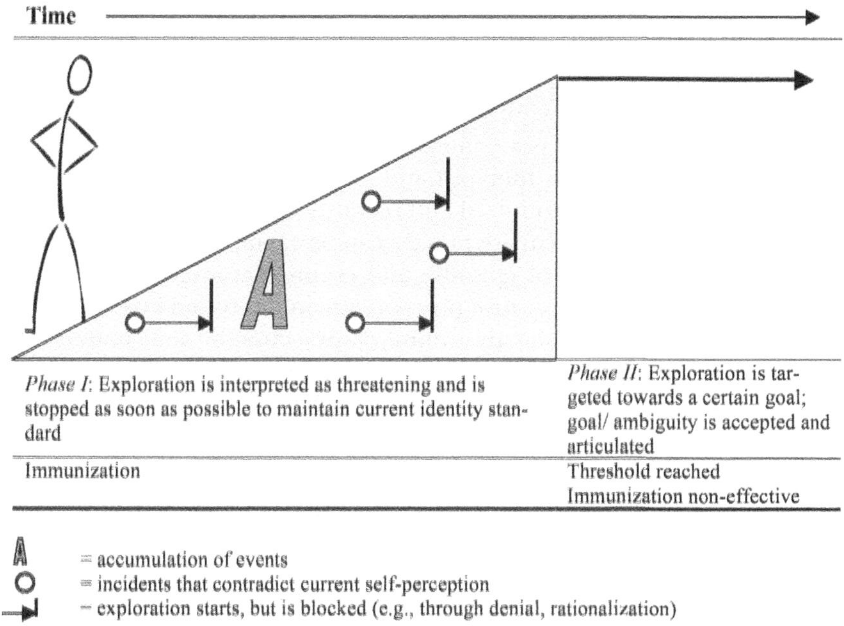

Figure 6.2 Different phases of identity formation.

about homosexual content, and so on. The first time one feels something for someone of the same sex might be easily dismissed (Immunization: "Everyone dreams stuff like this, this does not mean I am gay!", cf. Greve & Wentura, 2003). Then the second time, one might get more suspicious, because it becomes less likely that it was only an aberration. Eventually, when one cannot ignore or make excuses for one's own feelings and thoughts, one has to accept the fact that the standard is not appropriate. If this is the case, exploration becomes goal-oriented, testing a possible new identity facet/standard, whereas before exploration was aimed to find out what this identity standard might be. In Phase II, events in Phase I are likely to be reinterpreted. For example, a feeling that was assigned to a developmental stage and thus not taken seriously in the past, now turns into a precursor of the new identity standard (Josselson, 2009). After the second exploration phase, the individual is likely to commit to a certain identity standard ("I am homosexual!").

SARAH: IDENTITY EXPLORATION IN PSYCHOTHERAPY AS A RECURSIVE PROCESS (CANADA)

We look now into psychotherapy as a formal context where identity can be the object of analysis and reflection in order to create changes. Clients often present themselves seeking professional help in the midst of an identity crisis or a self-rupture: They are no longer at peace with who they are, yet they are still struggling to understand what to do and who to be (incongruence and discrepancies are sensed at an unbearable level).

Hence, psychotherapy is a context where exploration is explicitly elicited and engaged by both therapist and client, where sometimes identity structure is recursively scrutinized and reconstructed in order to achieve a new organization. This recursive process can be exemplified in this specific context through the use of episodes and events that are sometimes told and retold in the communicative process with an interested interlocutor—the therapist—both as a way to ground, "work though," and understand our present selves and to depart from them creating new understandings, meaning bridges (Brinegar et al., 2006), and alternative forms of self-integration or find a "new piloting system" (Zittoun, 2006, p. 191).

Our following example is Sarah[14] (pseudonym), a German immigrant woman in Canada in the beginning of the 1990s who participated in the York I Depression Study (cf. Greenberg & Watson, 1998). This study was a comparative clinical trial that contrasted the outcomes of different psychotherapeutic modalities in the treatment of depressive clients. This participant was a 35-year-old woman recently divorced at the time, showing symptoms of clinical depression at the onset of treatment. She attended a total of 18 sessions of individual emotion-focused therapy and was considered a good-outcome case at the end of treatment due to a significant reduction in the depressive symptoms exhibited in the beginning (for further details, cf. Honos-Web, Stiles, & Greenberg, 2003, or Greenberg & Watson, 1998). Sarah entered therapy seeking for professional help to address her increasing social isolation and interpersonal barriers and to work on her difficulties in expressing herself and being assertive towards others, as she usually put others' needs first by adopting a maladaptive caretaking attitude (Cunha et al., in press; Honos-Web et al., 2003). Sarah's therapist was a female doctoral student also in her thirties with prior clinical experience and training either in client-centred therapy and emotion-focused therapy (Honos-Web et al., 2003).

We have selected three excerpts of the therapeutic conversation in three different sessions of Sarah's case that will allow us to discuss the role of recursivity in a progressive awareness and elaboration of meaning of past events that are revisited, re-experienced and reconstructed in the present in order to achieve new understandings and insight into the current identity crisis.

In these excerpts Sarah is reflecting upon her past trying to find and explore the origins and reasons of her present difficulties—her barriers and excessive caretaking attitude—in her developmental history with her family. The escalation of the exploration process that these excerpts present culminates with the client's insight and coordinated definition (of therapist and client) of a *"people pleasing"* pattern as a problematic, constant theme in Sarah's life.

Again, we will not be faced with a straightforward depiction of events—the actual truth of what happened—but the meanings that are attributed to memories, in the present act of interpretation of past events that are always reconstructed by the self in the here and now of this specific dialogical encounter.

Session 1[15]

Therapist: Well the reason I ask is it's still sort of hard to sometimes explain to people what really happens [in psychotherapy] and there's a lot of discussion of how exactly it works but my feeling is that you get inside and you check out first of all "what am I feeling?" and then by understanding what's going on, you understand what you need to do to change that.

Sarah: Mm-hm. Yeah because there has been a lot of different things. . . . I remember when I was growing up, [T: mm-hm.] you know, *just little things people said to me and I guess it really affected me heavily*. . . . I guess one of the awful ones was: "kids don't speak or interrupt when adults are talking to each other!" . . . or actually pretending to my family, my parents, my grandparents, *that I am enjoying a gift but I don't*! . . . [short laugh]

T: So you remember even back then doing these things.

S: Oh yes. *It goes back a long, long way*, always doing this and feeling well, I'm *not really showing how I feel* about this . . .

T: How you really feel, yeah.

S: And *feeling responsible also for making them feel good.*

T: Yeah, *if they brought you a gift you want them to feel that was a good idea and that you're happy.*

In this excerpt from the first session, therapist and client are trying to understand each other's roles and goals in psychotherapy: The therapist is trying to present the potentialities of the therapeutic process in order for the client to understand how she can benefit from it (explore her own feelings and realize what she needs and wants to change), and the client is contextualizing her difficulties in her life-story. This triggers Sarah to loop into former events, with increased self-reflexivity, as she shares remembering growing up with certain powerful messages from others inhibiting genuine self-expression (*acting as if enjoying a gift*) and being guided towards an orientation to focusing on others' needs while devaluing her own (*feeling responsible for making others happy*). This structures an interactional process that is frequently activated in therapy from this point on: reflecting on how-I-am-now (*not really showing how I feel*) in the presence of the therapist as interlocutor which, in return, reflects and mirrors me to myself (*feeling responsible also for making them feel good*).

The peripheral example of this caretaking and self-dismissive attitude, like falsely pretending to enjoy a gift in the last excerpt from first session, is spontaneously recollected and revisited by the client in Session 7; this time retold and amplified in the therapeutic conversation and building on past efforts to understand in therapy what these events mean to Sarah and about Sarah (Session 1), as the therapist orients the client into a wider exploration of her past and its meanings.

Session 7: Recalling the Episode of the Doll House Gift

> **T:** *You said it feels like it all started in your family, like somehow just sensing that's what you needed to do to survive almost.*
>
> **C:** Yeah, well *it started at a really early age* and I remember a couple of incidences when I was around three or so, and the thing was like Christmas was coming up and I wanted a farmhouse and animals for Christmas and instead I walk into the living room with the Christmas tree and what do I find? A doll house and it was . . . *I still recall the instance and how disappointed I was about that, but I knew at the same time that everybody expected me like to be thrilled about it*, with the doll house. And I remember: okay, just do it!, you know, and it was just "Oh yeah, I really like it, and thank you . . ."
>
> **T:** So you almost remember that real strong early example of how you really felt and what that meant to them buying you the doll house instead of the farmhouse and *covering that up and saying "I'll do what I'm supposed to do, I'll make everyone happy."*
>
> **C:** Yeah and *I just cannot figure out how this all came about. . . .* You know, I remember this from three years old and what was preceding that or what are the circumstances which led to that, I have just absolutely no idea. . . . And it *scares me to remember that it goes back that far and that I have carried it through a long time.*
>
> **T:** If you think of that instant, do you feel that you told people about wanting a farmhouse and that you expected you might get that?

C: Well, I think I put it forward because they were asking about it, and then yeah I really expected to walk in and see the farmhouse with the animals.

T: Somehow you really expected it and then you saw the doll house.

C: It was like "No, this is not what I wanted..."

T: And *what does it mean to you that it was a doll house? What did you make of that? It seems like it was more than just "oh well they made mistake"...*

C: Well, *it was like they didn't listen to what I really wanted.*

T: Somehow *the sense of not being heard,* no one caring about what you really wanted.

Again here, an early event from the past is seen through a template that has organized Sarah's identity: focusing on others instead of expressing herself spontaneously and genuinely. In this session, the recollection of this particular episode (inner reaction and presentation of the self when receiving a gift) is expanded and built upon the initial efforts captured in Session 1—this process illustrates nicely the interplay between identity process and structure that can happen while we reflect recurrently in a given event. As the therapist starts framing her present problems as consequences of an adaptive process developed throughout her past, the general rules that underlie her problems start becoming differentiated, and certain interactional patterns become evident to the reflecting self. According to this observation standpoint that is enhanced by the therapist, a recollection of the past is brought to the foreground, being narrated by the client and elaborated in the here and now: the episode of the switched gift. This event is being understood and experienced now in a different way than it probably was then: novelty is being created from the recursive process. The client is adopting the wider view of a narrator, an observer, distancing herself from the mere perspective of an actor of the event. A discrepancy in the self triggers this exploration process: From Sarah's point of view now, it is no longer more important to keep others happy at the cost of inhibiting herself. Now, this observer perspective—of a narrator of her own story—is activated and it enables her to recognize a pattern in her life: *It scares me to remember that it goes back that far and that I have carried it through such a long time.* Sarah is carrying out an explicit exploration of her past with the present interlocutor and rejecting this usual way of being and acting, whereas before, the exploration process might have been only implicit or unintentional and encapsulated within the self. Through the dialogue with her therapist, Sarah is able to revisit the event and explore it, and novelty emerges. A self-reflective and more encompassing view is being promoted and adopted first by the therapist and then mimicked and developed by the client, facilitating her awareness and discovering novelty in the past. Through the progressive looping back into the past, some events become more differentiated and clearer now to their narrator, adopting an observer position. We consider novelty to be, in this case, the insight and recognition of previously uncovered automatic, abbreviated interactional processes that now appear in full view to the observing self as usual processes in Sarah's relationships. The therapist introduces more potential development in the therapeutic dialogue: *What does it mean to you that it was a doll house? What did you make of that? It seems like it was more than just "oh well they made mistake. . . . "* Through these attempts to speak for the client, the therapist is actually tentatively providing tools to foster further exploration and understanding in the client. Sarah's answer, *It was like they didn't listen to what I really wanted,* goes further in the recognition of the links between her present difficulties and former interpersonal enactments and is mirrored and completed by the therapist's following turn, restating back to Sarah: a pervasive *sense of not being heard* by others. This dialogue illustrates here, very clearly, how the (re)constructive process of identity exploration and revision is actually a product of co-construction in the therapeutic dialogue.

In the following, Session 8, the observation position continues to be dyadically amplified. The following excerpt presents another recollection—or, better phrased, the re-visiting and reinterpretation of a past event that, through this progressive process of self-and other-reflexivity, allows the depiction and clarification of the problematic theme in the participant's own words: *people pleasing.*

Session 8: Recalling an Early Attunement to Others' Feelings

> **C:** Yeah, because like the earliest memories I have, I guess preceding age four, *they're always in regards like to pleasing people, like to make them feel good. . . .* Well, I must've been—I don't know, two-and-a-half or so—and I was still, my mom still brought me the baby bottle in the morning with, I don't know, milk or whatever, and I have this one memory [laughs] that she came in... well, it"s really different how all this worked . . . I was sharing a room with my great aunt and we were all living—like my grandparents, my great aunt, my parents, and me—we were living together in the house and my mom was working during the day but in the mornings she always would bring in this baby bottle for me, and I remember one morning she brought it and I took it and I started sucking it "oh god, I hate this taste!" [laughs] but somehow

I knew it is important for her, you know, like part of her daily routine and also the contact between me and her and the preparations and . . . I couldn't say anything [laughs].

T: So it's pretty amazing at two-and-a half. . . . Most kids would just throw the bottle and say "I hate this" and *you were already so attuned that this would hurt her* [C: Yeah.] that I'm going to drink this anyway and not ruin it for her.

C: Yeah and that has carried on for a very long time that I just knew that it was important to people, to the family, to my parents and my grandparents, my great aunt, to do things like that. . . . Like they had spent time and thought on it, to get certain things or whatever it was, and I couldn't say "No, I don't like it, it's not what I want . . ." like, for instance, this doll house and [T: Yeah.] and farm house . . . [slight laugh]

T: Yeah. So just *always worrying not to hurt anyone's feelings* or . . .

C: Hmm . . . [pauses] Well, it's so back then, like I was kind of like the princess, being the oldest grandchild and the first child. And I know that they just really adored me and loved me and I really didn't want to disappoint them, and *somehow all this kind of snowballed* [T: Hmm.] that I carried that over to other people as well, and it is really difficult to turn around. . . .

T: Yeah, to deal with it . . . And yet it sounds like for you *this was sort of an example or scenario that really was a theme of how you live your life,* kind of always being very concerned not to hurt anyone's feelings, not to [C: Yeah.] rock the boat . . .

As the explorative process keeps on going like we notice in this excerpt, Sarah and her therapist start labeling the maladaptive interpersonal pattern as "people pleasing." This triggers another looping back into the client's past in the recognition of another illustrative event of this coherence of past-to-present interpersonal process: recalling the routine of the morning baby bottle. The meaning construction that Sarah presents of this precocious episode in her life illustrates the outcome of her identity exploration: This episode—either a real or fabricated memory—is understood now as an example of Sarah's sharp attunement to other's needs from an early age on. We highlight, again, that the important thing in our view is not the accuracy of the experience, but rather the meaning that is being assigned to the episode, presently being looked at and dynamically reorganized in a different way than before. This concern with others' feelings, as Sarah now argues, has snowballed into a generalized interpersonal pattern that is considered maladaptive in light of Sarah's present view upon herself and current values and goals. This event, like her therapist then reflects (*this was sort of an example or scenario that really was a theme of how you live your life*), becomes integrated in her identity structure as something that is very characteristic of a problematic life theme that Sarah eagerly wants to abandon and change, no longer being valued by the client. In this sense, it becomes a meaning bridge that establishes the need for explicit and intentional identity changes, providing a sense of continuity within an identity rupture.

CONCLUSIONS

We demonstrated that recursion is part of the exploration that is taking place when identities develop. These processes are very different in nature, but they also show basic principles:

- *Exploration is ongoing, as a continuous interpretation of the self in an attempt to create stability and coherence in flux.* Once an event causes some identity rupture and is understood as highly relevant to the self, it is interpreted in a way that creates continuity and meaning. However, it can also be continuously revisited and built upon. It can be given new meanings as the present context and developmental stage changes and personal stories get re-embedded afresh to meet new circumstances; new meanings yet also related to previous interpretations. Certain open-ended thoughts/discrepancies/amorphous feelings that are not full-blown exploration at the time may be retroactively recognized as such. These moments might retroactively be turned into crucial moments (as meaning bridges)—but they also may not.
- *Full-blown crises are no imperatives for exploration.* Sometimes past events—of which the recognition was blocked or which were interpreted in a certain way to avoid incongruence—"accumulate" (see Figure 6.2; examples for sexual identity development), or are reinterpreted later (see Miri and Sarah). Individuals make sense of the present by revisiting past events, by looping back. Exploration can thus be understood as a collaborative process between a "now person" and a "then experience": It is happening now and is happening then.
- In studies focusing on exploration and commitment, exploration is often described as a person thinking of the future and examining who he is now in relation to the future. *We have shown that people look backward to the past as an act of exploration.* Exploration needs to go back to the past. In order to make sense of choices for the future, the past needs to be reorganized to support the paths chosen. This should be more explicitly addressed in studies examining exploration processes.

Figure 6.3 Looking back at previous identities created by looking back.

- *We add to previous conceptions of identity such as the ICT model.* In the ICT model a person compares a current self with new feedback—perhaps changing the current perception (Figure 6.3). We show how the current perception is anchored in the past and in past interpretations of an even further past. So looking at who I am now might necessitate going back to the past and re-examine previous instances of exploration and tentative commitments.

On the basis of these two observations, we also argue that recursive movement of looking to the past, revisiting and reconstructing it in light of our new present purposes, paralleled and balanced with an exploration of a future state can be crucial in therapy. In our view, it is the increasing awareness and grounding of the past events that presents the potential to launch a new/changed identity and perform new ways of relating to oneself.

The concept of recursivity seems to be a useful concept to elaborate further on identity development processes that are not as straightforward as sometimes assumed. Microgenetic analyses could profit from having a closer look at the different ways with which individuals recursively make sense of their lives by connecting past, present and future.

NOTES

1. Bartlett (1951) comes to the same conclusion when he writes about the *thinking mind.* He says that mental processes always link situations in some way by, for example, drawing from information available to solve new problems. Nev-

ertheless, also behavior can change; in Figure 1 A should then be replaced by A′ (variation of A) or different letter (new behavior).

2. What does not change is the human social construction (meaning). We still classify the observed action as push-up, no matter if it is the first or 100th push-up.
3. The push-ups will also not be exactly the same because of the strain put on the body, but the person will at least try to perform the same motion-sequence again and again (behavioral aspect), even if the inner monologue changes from "That was easy!" to "You can do one more, come on!" (cognitive aspect).
4. The person is asked to describe her-/himself, usually in the present, but sometimes also how she used to be, and how she would like to be in the future. When using questionnaires, change is captured by comparing the person's evaluations *of* (cross-sectional) or *at* (longitudinal) different points in time. The results of the according data analyses are statements whether change has occurred or not. In addition, the kind of change can be described (A is now more outgoing than A used to be). What is missing is the description of the process: B might also be more outgoing than B used to be, but whereas A changed due to a school year abroad (and all the experience that came with it), B had chosen to participate in self-confidence training.
5. Already Erik Erikson (1980, p. 22) had stressed that "the conscious feeling of having a *personal identity* is based on two simultaneous observations: the immediate perception of one's selfsameness and continuity in time; and the simultaneous perception of the fact that others recognize one's sameness and continuity."
6. For a critical examination of the construct *life phase*, see Janssen (2009).
7. Commitment stresses that choices come with certain obligations. The concept commitment has, yet, been redefined by other authors that stress the value commitment has for self-esteem (Meeus & Decovic, 1995).
8. In a study by Kidwell, Dunham, Bacho, Pastorino, and Portes (1995), 39% of the variance of adolescent exploration could, nevertheless, be explained by a factor named Identity Exploration Crisis, which was characterized by self-doubt, confusion, disturbed thinking, reduced ego-strength, and so on.
9. For a critical assessment of the identity status model see, for example, Meeus (1996) or Schwartz (2001).
10. Due to changes in the environment and in the frequency with which certain identity statuses could be observed, Marcia (1989) stressed the necessity to differentiate diffusion further. He postulates that there are four types: disturbed (maladaptive), carefree, developmental, and culturally adaptive diffusion. Disturbed diffusion has some overlay with the concept of ruminative exploration.
11. Grotevant (1987) talked about *processes* rather than *factors*. Bosma and Kunnen (2001) pointed out though that only (b) and (e) are processes; all other elements have to be considered factors.
12. Even maleness and femaleness are not distinct categories, but should rather be seen as the end points of a continuum (cp. for example Watzlawik, 2004a).
13. Miri is quoting (with some changes) from the Mishnah, *Tractate of the Fathers*, 6,2.

14. We are very grateful to Lynne Angus and Leslie S. Greenberg from York University (Toronto, Canada) for allowing us to analyze the transcripts of the Sarah case.
15. These transcripts were edited in order to eliminate redundancies and specificities of oral speech, although maintaining the content.

REFERENCES

Bamberg, M. (2010). Who am I? Narration and its contribution to self and identity. *Theory & Psychology, 21*(1), 1–22.

Bartlett, F. C. (1951). *The mind at work and play.* London, England: George Allen & Unwin.

Berman, A. M., Schwartz, S .J., Kurtines, W. M., & Berman, S. L. (2001). The process of exploration in identity formation: The role of style and competence. *Journal of Adolescence, 24,* 513–528

Berzonsky, M. D. (1989). Identity style: conceptualization and measurement. *Journal of Adolescent Research, 4,* 267–281.

Bowlby, J. (1969). *Attachment and loss: Attachment.* New York, NY: Basic Books.

Bosma, H. A., & Kunnen, E. S. (2001). Determinants and mechanisms in ego identity development: A review and synthesis. *Developmental Review, 21,* 39–66.

Brinegar, M. G., Salvi, L. M., Stiles, W. B., & Greenberg, L. S. (2006). Building a meaning bridge: Therapeutic progress from problem formulation to understanding. *Journal of Counseling Psychology, 53,* 165–180.

Clifford, P. A., Friesen, S., & Jardine, D. (2001). *The ontology of hope: Lessons from a child.* Paper presented at the 82nd Annual Meeting of the American Educational Research Association, Seattle, WA. Retrieved from http://jan.ucc.nau.edu/ ~chaplx-p/docs/ClifordetalAERA2001Paper.pdf

Cunha, C., Gonçalves, M. M., Valsiner, J., Mendes, I., & Ribeiro, A. P. (2013). Rehearsing renewal of identity: Re-conceptualization on the move. In M. C. Bertau, M. M. Gonçalves & P. Raggatt (Eds.), *Dialogic formations: Investigations into the origins and development of the dialogical self* (pp. 205–233). Charlotte, NC: Information Age.

Deci, E. L., & Ryan, R. M. (2000). The darker and brighter sides of human existence: Basic psychological needs as a unifying concept. *Psychological Inquiry, 11,* 319 338.

Driesch, H. (1925). *The crisis in psychology.* Princeton, NJ: Princeton University Press.

Erikson, E. H. (1968). *Identity: Youth and crisis.* New York, NY: Norton.

Erikson, E. H. (1980). *Identity and the life cycle.* New York, NY: Norton.

Flum, H., & Kaplan, A. (2006). Exploratory Orientation as an Educational Goal. *Educational Psychologist, 41*(2), 99–110.

Freud, A. (1937). T*he ego and the mechanisms of defence.* London, UK: Hogarth Press.

Greenberg, L. S., & Watson, J. C. (1998). Experiential therapy for depression: Differential effects of client-centered relationship conditions and active experiential interventions. *Journal of Psychotherapy Research, 8,* 210–214.

Greve, W., & Wentura, D. (2003). Immunizing the self: Self-concept stabilization through reality-adaptive self-definitions. *Personality & Social Psychology Bulletin, 29*(1), 39–50.

Grotevant, H. D. (1987). Toward a process model of identity formation. *Journal of Adolescent Research, 2,* 203–222.

Grotevant, H., & Cooper, C. (1986). Individuation in family relationships. *Human Development, 29,* 82–100.

Hall, S., & du Gay, P. (1996). *Questions of cultural identity.* Thousand Oaks, CA: Sage.

Honos-Webb, L., Stiles, W. B., & Greenberg, L. S. (2003). A method for rating assimilation in psychotherapy based on markers of change. *Journal of Counseling Psychology, 50*(3), 189–198.

Janssen, D. F. (2009). Life course staging as cultural and subjective practice: Review, critique, and theoretical possibilities. *Culture and Psychology, 15*(4), 541–560.

Jenkins, R. (2008). *Social identity* (3rd ed.). New York, NY: Routledge.

Josselson, R. (2009). The present of the past: Dialogues with memory over time. *Journal of Personality, 77*(3), 647–668.

Kerpelman, J. L. (2001). Identity control theory, exploration, and choice: A commentary on Schwartz's "The evolution of Eriksonian and Neo-Eriksonian identity theory and research." *Identity, 1*(1), 81–86.

Kerpelman, J. L., Pittman, J. F., & Lamke, L. K. (1997). Toward a microprocess perspective on adolescent identity development: An identity control theory approach. *Journal of Adolescent Research, 12*(3), 325–346.

Kidwell, J. S., Dunham, R. M., Bacho, R. A., Pastorino, E., & Portes, P. R. (1995). Adolescent identity exploration: A test of Erikson's theory of transitional crisis. *Adolescence, 30,* 785–793.

Kraus, W. (2000). Making identities talk: On qualitative methods in a longitudinal study [online journal]. *Forum: Qualitative Social Research, 1*(2). Retrieved from http://www.qualitative-research.net/index.php/fqs/article/viewArticle/1084/2367

Kroger, J. (2006). *Identity development: Adolescence through adulthood.* Thousand Oaks, CA: Sage.

Luyckx, K., Goossens, L., Soenens, B., & Beyers, W. (2006). Unpacking commitment and exploration: Preliminary validation of an integrative model of late adolescent identity formation. *Journal of Adolescence, 29,* 361–378.

Luyckx, L., Schwartz, S. J., Berzonsky, M. D., Soenens, B., Vansteenkiste, M., Smits, I., & Goossens, L. (2008). Capturing ruminative exploration: Extending the four-dimensional model of identity formation in late adolescence. *Journal of Research in Personality, 42,* 58–82.

Maccoby, E. E. (1999). *The two sexes: Growing up apart, coming together.* Cambridge, MA: Harvard University Press.

Marcia, J. E. (1966). Development and validation of ego-identity status. *Journal of Personality and Social Psychology, 3,* 551–558.

Marcia, J. E. (1980). Identity in adolescence. In J. Adelson (Ed.), *Handbook of adolescent psychology* (pp. 159–187). New York, NY: Wiley.

Marcia, J. E. (1987). The identity status approach to the study of ego identity development. In T. Honess & K. Yardley (Eds.), *Self and identity* (pp. 161–171). London, UK: Routledge & Kegan.

Marcia, J. E. (1989). Identity diffusion differentiated. In M.A. Luszcz & T. Nettelbeck (Eds.), *Psychological development: Perspectives across the life-span* (pp. 289-295). North-Holland: Elsevier.

Marcia, J. E. (1993). The ego identity status approach to ego identity. In J. E. Marcia, A. S. Waterman, D. R. Matteson, S. L. Archer, & J. L. Orlowsky (Eds.), *Ego identity* (pp. 22–41). New York, NY: Springer.

Marcia, J. E. (2007). Theory and measure: The identity status interview. In M. Watzlawik & A. Born (Eds.), *Capturing identity: Qualitative and quantitative methods* (pp. 1–14). Lanham, MD: University Press of America.

Marcia, J. E., & Archer, S. L. (1993). Identity status in late adolescence: Scoring criteria. In J. E. Marcia, A. S. Waterman, D. R. Matteson, S. L. Archer, & J. L. Orlofsky (Eds.), *Ego identity* (pp. 205–240). New York, NY: Springer.

Meeus, W. (1996). Towards a psychological analysis of adolescent identity: An evaluation of the Epigenetic Theory (Erikson) and the Identity Status Model (Marcia). In K. Hurrelmann & S. F. Hamilton (Eds.), *Social problems and social contexts in adolescence* (pp. 83–105). New York, NY: de Gruyter.

Meeus, W., & Dekovic, M. (1995). Identity development, parental and peer support in adolescence: Results of a Dutch national survey. *Adolescence, 30,* 931–944.

Meeus, W., Iedema, J., & Maassen, G. H. (2002). Commitment and exploration as mechanisms of identity formation. *Psychological Reports, 90,* 771–785.

Offer, D. (1969). *The psychological world oft he teenager.* New York, NY: Basic Books.

Perret-Clermont, A.-N. & Zittoun, T. (2002). Esquisse d'une psychologie de la transition [Sketch for a psychology of transitions]. Education permanente. *Revue Suisse pour la Formation Continue, 1,* 12–15.

Sarbin, T. R. (1983). Place identity as a component of self: An addendum. *Journal of Environmental Psychology, 3*(4), 337–342.

Schachter, E. P. (2002). Identity constraints: The perceived structural requirements of a 'good' identity. *Human Development, 45*(6), 416–433.

Schupp, K. (1999). "Sie liebt sie"—"Er liebt ihn": Eine Befragung von lesbischen, schwulen und bisexuellen Jugendlichen ["She loves her"—"He loves him": Survey among lesbian, gay, and bisexual adolescents]. In T. Hofsäss (Ed.), *Jugendhilfe und gleichgeschlechtliche Orientierung* [Youth aid and same-sex orientation] (pp. 21–44). Berlin, Germany: Verlag für Wissenschaft und Bildung.

Schwartz, S. J. (2001). The evolution of Eriksonian and Neo-Eriksonian identity theory and research: A review and integration. *Identity, 1*(1), 7–58.

Smith, A. D. (1993). *National identity.* London, England: Penguin Books.

Stephen, J., Fraser, E., & Marcia, J. E. (1992). Moratorium-achievement (MAMA) cycles in lifespan identity development: Value orientations and reasoning system correlates. *Journal of Adolescence, 15,* 283–300.

Vaillant, G. E. (1992). *Ego-mechanism of defence: A guide for clinicians and researchers.* Washington, DC: American Psychiatric Press.

Watzlawik, M. (2004a). *Uferlos? Jugendliche erleben sexuelle Orientierungen* [Shoreless? Adolescents experience sexual orientations]. Norderstedt, Germany: Books on Demand.

Watzlawik, M. (2004b). Experiencing sexual orientation: A comparison between American and German adolescents. *Identity, 4*(2), 173–189.

Watzlawik, M., & Heine, N. (Eds.). (2009). *Sexuelle Orientierungen. Weg vom Denken in Schubladen* [Sexual Orientations. Thinking outside the box]. Göttingen, Germany,: Vandenhoeck & Ruprecht.

Yang Costello, C. (2005). *Professional identity crisis: Race, class, gender, and success at professional schools.* Nashville, TN: Vanderbilt University Press.

Zittoun, T. (2006). Dynamics and inferiority: Ruptures and transitions in self development. In L. Mathias Simão & J. Valsiner (Eds.), *Otherness in question: Labyrinths of the self* (pp. 187–214). Charlotte, NC: Information Age.

CHAPTER 7

SAINT IGNATIUS OF LOYOLA'S SPIRITUAL EXERCISES AS A RECURSIVE EXPERIENCE

Luis Roberto Rivera, Ana Victoria Prados, Sandra Liliana Londoño, and Mauricio José Cortés
Pontificia Universidad Javeriana of Cali in Colombia

INTRODUCTION

This chapter reflects upon the presence of recursion in Saint Ignatius of Loyola's *Spiritual Exercises.* Two particular works are the basis for this paper. The first one belongs to the Jesuit Xavier Melloni in his book *Mystagogy of Spiritual Exercises* (2001). The second one is by linguist Roland Barthes in his book *Sade, Fourier, Loyola* (1997). These authors approach the study of said exercises with different academic interests. Melloni (2001) analyzes them as an experience of mystical initiation, Barthes (1997) as an experience of language creation. The following is an interpretation of these studies from a perspective of recursion. This way, a new key in reading the Ignatian approach is proposed in which all elements within acquire a new meaning: notes, additions, exercises, exams, and so on, as will be developed further later on.

Cultural Psychology of Recursive Processes, pages 195–217

The recursion found in Spiritual Exercises is understood from the categories of autosimilarity and linguistic interaction. These categories are shown in the second section of the chapter, respectively. Based on them, a particular reading of the Ignatian text is developed that lets us exemplify the way spiritual exercises contain and take advantage of recursion. These developments are stated in section three. As a starting point to this chapter, we describe the objective and methodology of spiritual exercises.

This approach opens a new interpretative horizon for spiritual exercises. It allows connecting them with processes of self-care and identity development in which the same kind of recursion is present. Traditional mystic–ascetic approach can be surpassed. The comprehension of the Ignatian approach is promoted as a practice in constructing the subject in virtue of the novelty required by the linguistic recursion.

On the other hand, the aspects involved in the recursion of the exercises enrich the comprehension of other practices oriented towards the construction of the subject. The set of elements that Saint Ignatius takes into account and the way he does so can be taken advantage of in other activities. For example, management of schedule, environment, isolation, and work scheme are concrete aspects that build recursion and establish the conditions of possibility for novelty to arise and the consequent transformation of the subject.

OBJECTIVE AND METHODOLOGY OF SPIRITUAL EXERCISES

This section presents summarily the experience of Saint Ignatius of Loyola's spiritual exercises. The objective is to outline the context in which the following sections can be comprehended, without expecting to be exhaustive in describing the Ignatian approach.

Spiritual exercises are an expression of the human effort in the search for meaning, in building an identity, in self-care. Saint Ignatius of Loyola presents their (spiritual exercises') objective in number [1] of his book when describing what's meant by such:

> For as strolling, walking and running are bodily exercises, so every way of preparing and disposing the soul to rid itself of all the disordered tendencies, and, after it is rid, to seek and find the Divine Will as to the management of one's life for the salvation of the soul, is called a Spiritual Exercise [1].

The expression "every way" indicates an open set of experiences described by intensional definition and not by extensional definition. The formula presents a path towards God's will with three parts: making oneself available, searching, and finding. Therefore a work is foreshadowed in which a decision must be made: one must decide (finding) in freedom (disposition) among several alternatives (search). This work is carried out in a dialogue with God.

The foregoing definition is directly in line with the ascetic approach to exercises. According to this way of seeing them, their objective would be the decision that he who exercises makes at the end of the second week. However, the comprehension of the exercises' reach requires taking in consideration the tension between this approach and a mystical reading of said exercises, according to which, the exercises are a path towards Christian sanctification, towards union with God. This tension disappears with an integrating vision postulating that a union with God is reached through choice. This is the Ignatian way of understanding the Christian mysticism.

It's also important to know the methodology of spiritual exercises. It's been established that the practices themselves aren't a closed set. On the contrary, it is a proposal of flexible and adaptable work to the exerciser. All in all, there are some directions that help to structure the experience, which are presented in St. Ignatius's book as *annotations, additions,* and *notes.*

Exercises are developed in about 30 days. It is recommended to carry them out in a place different from the exerciser's everyday life that allows interior and exterior silence. A spiritual director partakes in the experience, who will assign the exercises, adapting them however is considered necessary according to the experience that develops.

The Ignatian text comprises different instructions, admonitions, prayers, meditations, consciousness examinations, and norms, among others. They are divided in four "weeks" corresponding to the topics and not groups of seven days. In a typical day the exerciser does about five exercises, following the director's instructions, with the corresponding additions and examinations, and meets with the director as well to account for what's happened while doing them.

Exercises begin with the meditation called "Principle and Fundament" and then they continue with the first week's exercises, which aim to eliminate from the soul the deformity caused by sin: *deformata reformare.* During the second week, the reformed exerciser must conform to Christ; in other words, find the way to follow Him: *reformata conformare.* Upon the end of the second week, the exerciser makes a decision regarding it. During the third week, the purposes for joining Christ must ratify the decision: *conformata confirmare.* And lastly, in the fourth week comes the recognition that upon death, life begins through recognition of the experience of resurrection: *conformata transformare.* As a conclusion the Contemplation to reach Love is proposed. This aims for the person to discover how and how much God loves him (Order of Friars Minor, 1909).

The stages of spiritual life during Saint Ignatius' time were understood as being organized in three paths or stages: the purgative path, the illuminative path (the ascetic paths), and the unitive path (the mystic path) (Melloni, 2001). The purgative path refers to cleansing the soul of the attachments of sensitive desire. The illuminative path refers to raising the understanding toward God. The unitive way refers to purification of the will of joining God. These ways seem to be present during the first, second, and fourth weeks of exercises; but there is a fourth way, that of the third week, which is an innovation where the exerciser seems to freely choose upon contemplation of Christ's life and confirms the adopted decisions.

INTERNAL HOMOTHECY AND CASCADE IN THE SPIRITUAL EXERCISES

According to Melloni (2001), within exercises a fractal structure is found that favors the mystagogy of the spirit: that is to say, the deepening that human beings carry out in the spiritual mysteries from the Christian faith stand point. According to this author, Saint Ignatius designed a structure of reflection that obeys patterns of meaningful actions, organization of which may be represented with pentagons. Melloni identifies a pentagon for the level of exercise, another one for the workday, and another one for the full month. This way, he introduces the concept of homothecy in the structure of exercises and therefore that of recursion: The formal structure of the full experience keeps the same order as that of one day and this, in turn, the same as one hour of exercise. Melloni (2001) states that the pentagonal shape allows identification of the three paths of ascetic and Christian mysticism proposed by Jeronimo Nadal, who was contemporaneous of Saint Ignatius: the purgative path, the illuminative path and the unitive path. Figure 7.1 shows this structure of pentagonal homothecy.

In Melloni's opinion (2001), one of the ways of recursion identified in spiritual exercises is the property of being self-similar. This property is shared with the fractal figures. It is not precise to affirm that the structure of the exercises is a fractal, due to the fact that differences can be found between the definition of this type of figures and the way this property is identified in the mentioned spiritual experience. However, it is valid to affirm that the degree of self-similarity that the exercises possess is a way of recursion. To better understand the foregoing, it is necessary to define in more detail the recursion present in the fractals.

Mathematically speaking, the fractal formal definition exceeds the reach of this work, especially on the handling that implies the different concepts of dimension and differentiation. For the purposes of this chapter it suffices to say that fractals are autosimilar patterns: that is to say, figures representing

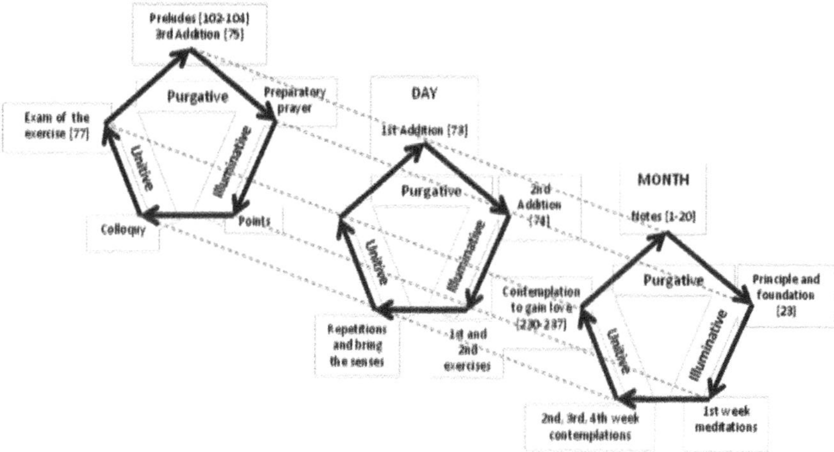

Figure 7.1 Graphical representation of the fractal structure of the exercises, according to Melloni (2001).

the same shape at any scale. In other words, the characteristics observed are the same both near and far away. It is a detailed pattern that repeats itself. This characteristic generates figures that can be divided in several parts that are a reduced copy of the whole. It is important to reiterate that autosimilarity is necessary but not enough to describe fractals (Mandelbrot, 1987).

On the other hand, fractal autosimilarity shouldn't necessarily have to be exact. That is to say, it is not required that at all levels of analysis identical duplicates are found. For this reason it is possible to talk about quasi-similarity without losing the fractal nature of the figure. This means that approximately the same patterns are found, but small differences may appear in different levels.

Moreover, some complex figures make problematic the quantitative description of autosimilarity, proper for the cases mentioned above. In many cases, the patterns may repeat randomly at different levels of approximation. In other words, they may appear in some cases with variations and in other cases not. In these figures, we speak of stochastic autosimilarity. Also, in other situations the quantitative approximation to describe self-similarity is not possible or is too complex, due to the nature or the shape of the fractal figure; however, autosimilarity is present and may be identified. The latter is referred to as qualitative autosimilarity.

Using a more formal language that allows both the quantitative and qualitative approaches, autosimilarity implies a mathematical type of recursion that requires two properties:

1. An initial case
2. A set of rules allowing the construction of a new case starting from the previous.

The set of rules referred to in the second point can be seen as a recursive algorithm, in the sense that each step is executed using the result of the same algorithm at the previous level of interaction.

The successions and series of numbers are an example of the application of the previous rules. A geometric succession is a sequence of numbers that complies with the property of maintaining a constant ratio between two successive numbers. Thus, it is possible to build the following geometric sequence using ½ ratio and initial number ½: ½, ¼, ⅛, $\frac{1}{16}$, and so on.

For the case of geometric fractals, the initial case is a geometrical shape, and the rules of construction imply the transformation of itself into a new autosimilar figure. For example, given the segment of a line in Figure 7.2a as initial case, Figure 7.2b is constructed substituting the central third with two segments of the same length as the one substituted, as shown. From this figure, it is possible to continue building the sequence applying the same procedure to each segment in Figure 7.2b. With iteration, Figure 7.2c is obtained. Then, Figure 7.2d and it can be continued infinitely. Using this recursion pattern, one of the most famous fractals was built: Koch's curve.

Additionally, two characteristics of fractal autosimilarity must be mentioned that, added to the previous ones, will allow the comprehension of their application for analyzing multiple natural phenomena, social phenomena and experiences, like spiritual exercises. Firstly, in some cases it requires more than one formation rule; in these cases the different rules are applied to different interactions. Secondly, their structure may be analyzed

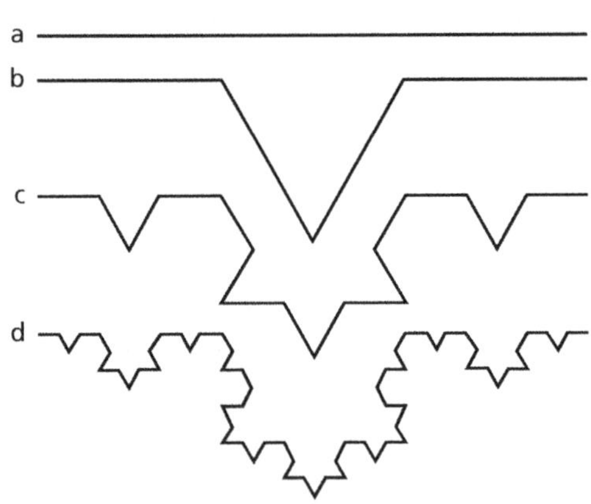

Figure 7.2 Example of the construction of a fractal figure.

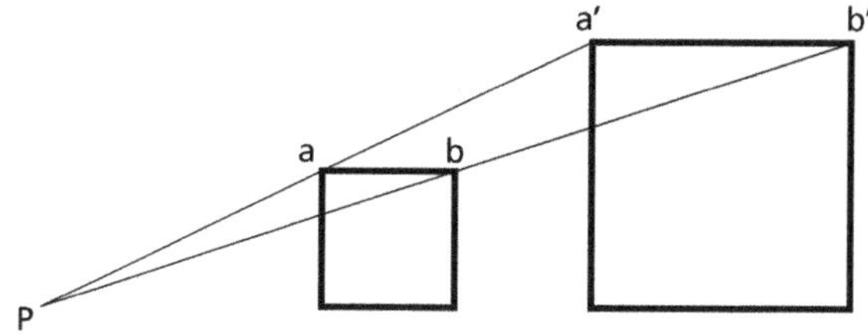

Figure 7.3 Example of homothecy.

at a large or small scale random play; that is to say, there are no limits to observe details of the figure; in other words, the recursion is infinite.

Geometrically speaking, autosimilarity is the result of the internal homothecy. This is a related transformation that projects each point of a region in space into another point that is itself at a certain distance from a point called center of homothecy, which is proportional to the distance of the initial point to said center of homothecy. The proportion is determined by a constant called ratio of homothecy. Figure 7.3 shows this scheme for a homothecy of center P and ratio K.

This type of transformation, applied to different parts of the figure, within different interactions of the recursive algorithm, build autosimilarity. It is important to clarify that for the case of fractals, it is not necessary to use only one point as center of homothecy, it isn't even needed to use the same ratio at each level of interaction.

The foregoing is part of the conceptual apparatus that formalizes the mathematical idea of recursion, which is the cornerstone of the proposal by Father Melloni (2001) on the fractal form of the spiritual exercise method.

Melloni (2001) presents autosimilarity of spiritual exercises as a condition of possibility for the mystagogical path these propose. As seen, the three paths of spiritual life suggest the structure of the exercise, of the day and the experience of the complete month. For the objectives intended by the author in his book, this level of analysis is sufficient. However, if the three paths scheme is abandoned, it is possible to identify more autosimilar elements in the exercises, as will be seen below.

This elaboration is obtained by separate analysis of the three elements of the Ignatian approach that Melloni (2001) casts as his scheme: order, method, and context. Method refers to how the exerciser operates; order refers to the sequence the experience follows; and context refers to the supports that facilitate it. Through the graphic in Figure 7.4, the elements involved in this approach of recursion can be visualized. It's an equilateral triangle, to reinforce the idea of equality in importance of the elements in its vertex.

The method by which exercises are carried out consists of a process constructed by four steps: feel, discern, decide, and confirm. In feeling, the

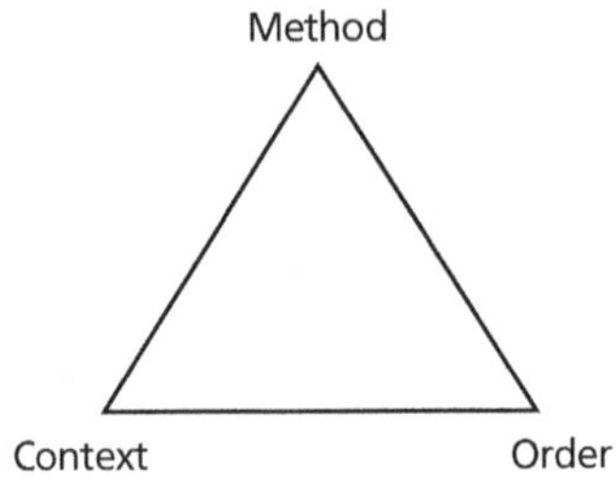

Figure 7.4 Autosimilar elements in spiritual exercises.

entire person experiences a situation; in the discerning step he comes back over what he has experienced to understand which elements come from God; in the deciding step he picks the elements coming from God to allow them to transform him; in the confirming step he makes sure the selected elements do come from God. Now then, confirming consists of a new process of feeling, discerning, and deciding in which the experience consists in the decision made. The four steps of the method are carried out through three different operations, which may be schematized, again, as an equilateral triangle (Figure 7.5).

On the other hand, exercises are carried out in a context that guides them or facilitates them. This context may be proposed to be composed by three elements: reference, one environment, and support. The reference consists of the exerciser's experience upon which the exercise is being projected—her life, the choice she expects to make. The environment is the circumstances in which the exerciser does her exercises; and finally, the support consists of different elements proposed by St. Ignatius to guide the experience.

Lastly, the order in accordance with Number 1 of the Exercises is proposed in three different steps: ordering life, seeking God's will, and finding it. Thus, we have two more equilateral triangles representing the aspects of the experience in Figures 7.6 and 7.7.

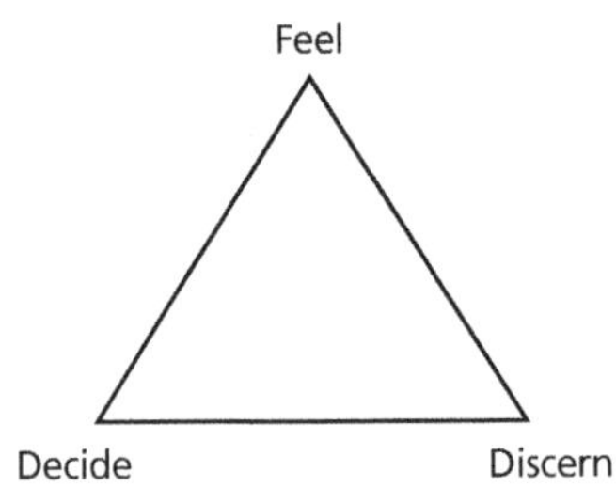

Figure 7.5 Methodological autosimilarity in spiritual exercises.

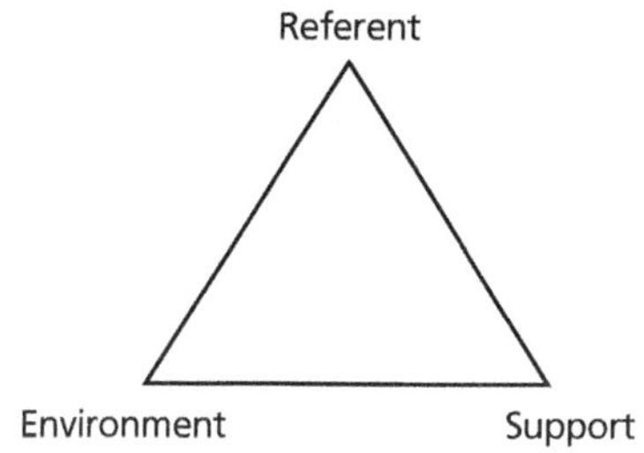

Figure 7.6 Contextual autosimilarity in spiritual exercises.

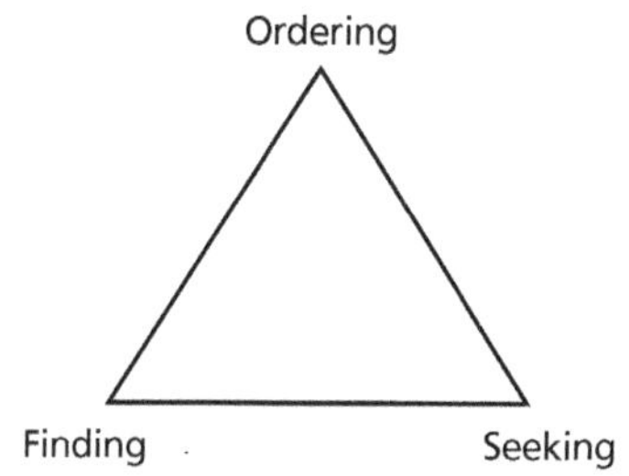

Figure 7.7 Auto-similarity in the order of spiritual exercises.

From the foregoing, it is observed that the way in which aspects from Figure 7.2 are represented is homothetic in the way each of its parts are represented. This is made possible due to the postulate of equality of importance of the nine identified elements. Thus, the following representation (Figure 7.8) may be proposed for the internal homothecy identified in *Spiritual Exercises*:

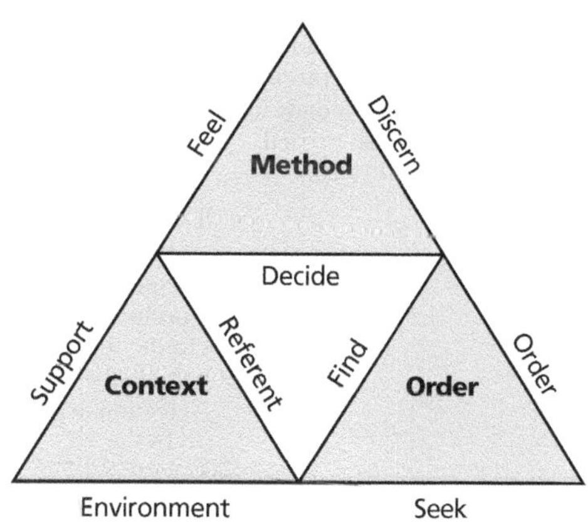

Figure 7.8 Model of internal homothecy for spiritual exercise.

Additionally, from the temporary view point, in spiritual exercises four levels can be distinguished clearly, these mark structures that begin and end: the hour of the exercise, the day, the week, the month. The first three are repeated various times and shape the corresponding upper level. It should be pointed out that Melloni's fractal model does not include the week level.

Saint Ignatius, in the book of exercises, submits the indications to proceed in each one of these levels not only for he who instructs but also for he who receives them. In the following tables, these indications are organized in the categories identified as homothetic. Table 7.1 revolves around the method, Table 7.2, the context, and Table 7.3, the order of the exercises

TABLE 7.1 Recursive Elements in the Method

Level	Exercise time[a]	Day[b]	Week	Month
Feeling	Examination of the exercise's points. "*to reflect in order to get profit*" [106]	Performance of the two different daily exercises at midnight and when wakening up in the morning [72]	Working on the selected exercises according to the weekly request [162]	Work on the exerciser's life and his election [106, 188]
Discerning	Identification of thoughts, wishes or feelings coming from God. This is carried out in the colloquy [53, 157] and the evaluation at the end of the exercise [77]. It's based on the rules for the discernment [313–336].	Two repetitions are carried out to note the aspects in which more spiritual pleasure has been had [62] and bringing about the senses on the day's exercises before dinner and in the afternoon [72]. A daily test is also carried out [90].	The daily examinations [90] And the interviews by which the exercises are handed out [4–10].	During the second week the process of elections is carried out [175–187]
Deciding			The achievement of the weekly request (last prelude). On the second week, the election as well [188].	One must choose and confirm said election [188].

[a] In the particular, general and daily exams; in the meditations and contemplations and in the choice.

[b] On a complete "typical" day.

TABLE 7.2 Recursive Elements in the Order

Level	Exercise	Day	Week	Month
Ordering	Additions 1, 2 and 3 [73, 74, and 75] Preparatory prayer [46] and preludes of each exercise [47–49]	Examination on additions [90] Additions 8 and 9 [80, 81]	Additions 6, 7 and 10 [78, 79 and 82] Gradual entrance to the first week. Resting before the second week.	First week
Seeking	Points of each exercise	First and second daily exercise.	All the weekly exercises.	Second week
Finding	Colloquium and examination	Repetitions and application of the senses	Examination and conversation with director	Second, third and fourth week

[a] This element is not shown in the Ignatian book but in the Exercises Directory. For example: (Lop Sebastià, 2000).

TABLE 7.3 Recursive Elements in the Context

Level	Exercise	Day	Week	Month
Referent	Work is done on the exerciser's life and beginning in the third week on his election as well [106, 188].	Work is done on exerciser's life from third week onwards on his election as well. [106, 188].	Work is done on exerciser's life and from the third week onwards on his election as well [106, 188]. The weekly petition (last prelude).	Work is done on exerciser's life and fro the third week onwards on his election additionally. [106, 188]. Principle and foundation for the first week [23]. Call of the temporal king for the second [91–100]. The election for the third and fourth week
Environment	The place, the time. Additions 1, 2, 3, 4 [73–76]	Time distribution [72].	Additions 6, 7, and 10 [78, 79, 82].	Additions 8 and 9. [80, 81]. Retreat, silence [20]
Support	The exercise's text	Interviews with the director [4–10]	Rules [313–370].	Notes [1–20] The exercise's book in general.

From this analysis, Melloni's (2001) proposal (Figure 7.1) can be reconstructed to extend the fractal representation of St. Ignatius' exercises. This is shown in the following figures. Figure 7.9 shows the fractal structure of the Exercises' method; Figure 7.10, the fractal structure of the contextual elements themselves; and Figure 7.11, the fractal structure of their order.

Finally, it is possible to put together the separate analysis in one fractal representation of the exercises' recursion. To do this, a known fractal

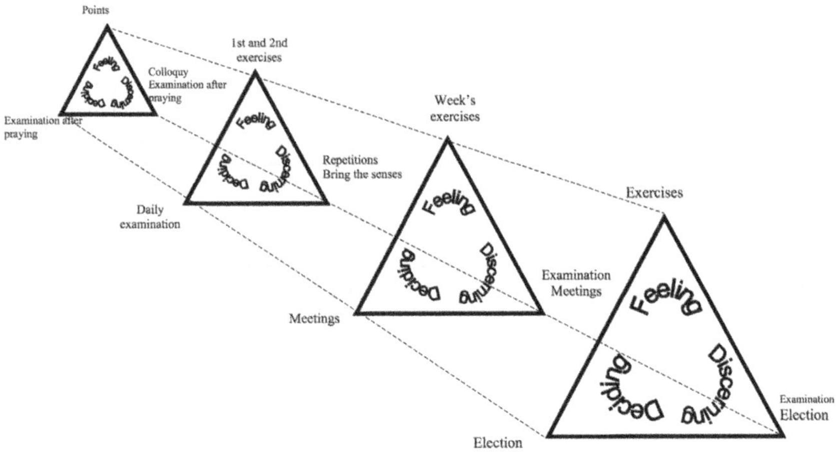

Figure 7.9 Fractal representation of the spiritual exercise method.

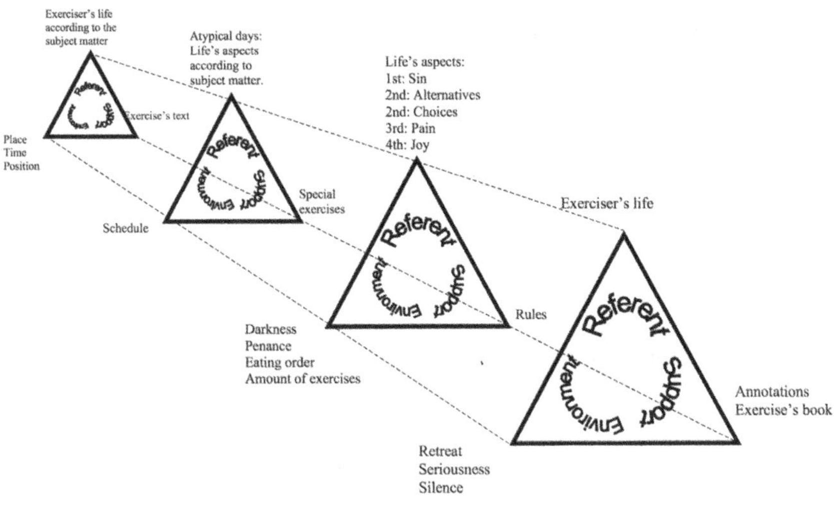

Figure 7.10 Fractal representation of the spiritual exercises fractal context.

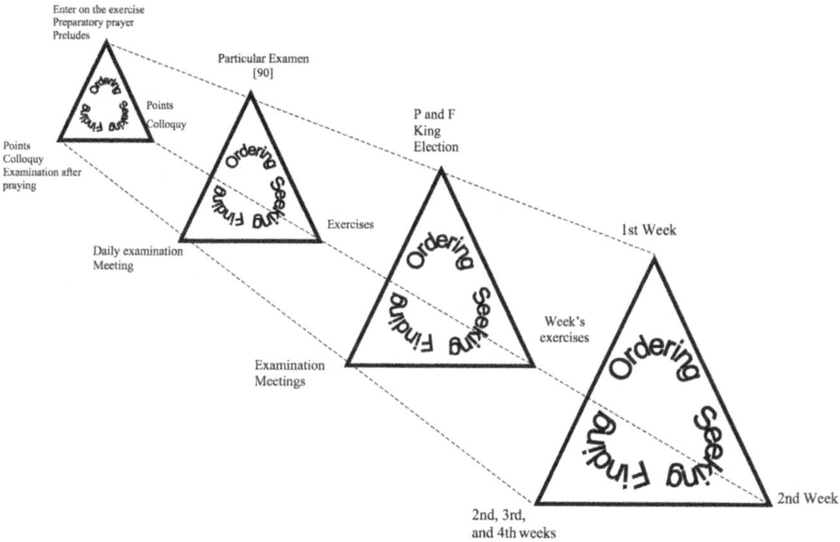

Figure 7.11 Fractal representation of the order in spiritual exercises.

is used, the Sierpinski triangle, introduced in 1974, of which the model shown in Figure 7.8 corresponds to the first level of iteration.

As it has been established, the exercises exhibit this model on four levels: The exercise's shape reproduces during the day, this one on the week and lastly this other on the month. The recursion presented in Tables 7.1, 7.2, and 7.3 is not evidenced as homogeneous. As can be seen, some cells of the tables are being left blank and others have their contents repeated. However, the fractal geometry assumes these irregularities and allows a unified representation of all complete exercises to be proposed.

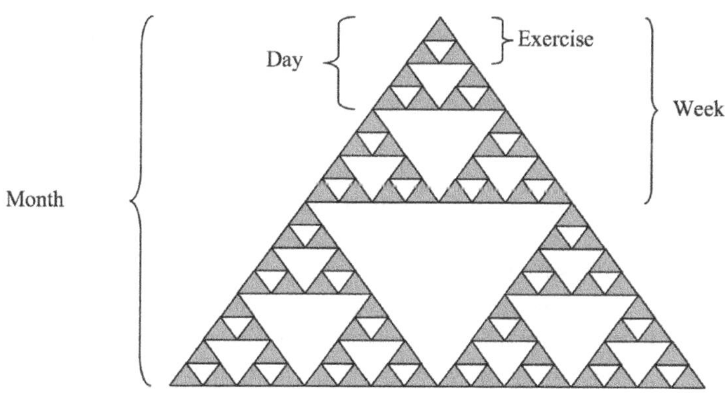

Figure 7.12 Fractal model for the complete experience of spiritual exercises.

Considering that some of the analyzed levels are contained within each other, the representation sought may be found through four iterations of the Sierpinsky triangle, one for each temporary level, as seen in Figure 7.12.

RECURSION IN LANGUAGE: LINGUISTIC RECURSION IN SPIRITUAL EXERCISES

Following Barthes (1997), *Spiritual Exercises* is a very particular text, designed as a code serving to decipher God's message to humankind. The novelty of the text consists in being a language, therefore, facilitating the dialogue between God and human beings. Lora-Garcés (2005), quoting Barthes, points out that this dialogical structure of the exercises is very original. It promotes a personal, unitive, and loving experience with God, since it introduces a dimension of interaction not found in other kinds of religious experiences.

Barthes (1997) explains that the four weeks can be understood as two moments, one before and one after, separated at the end of the second week. At this point, an empty space and a silence in the dialogue occur. This is the moment where the choice is located, the act of human freedom that confirms or not that which has been discovered about God's will for itself and which arouses multiple emotions. This empty space, Barthes assures, is extremely novel and makes meaning emerge.

There's controversy around the etymology of the word *dialogue*. Some point out that it is derived from the Greek word "*dialogos*" derived from the verb *dialegesthai*, which refers to the conversing, the verbal interaction between two or more participants (Linell, 2009): a relationship between communicative messages (Valsiner, 2007). Dialoguing is a typically human activity that implies the construction of meaning; in other words, it is a semiotic practice of action in interaction, thought, or communication.

Dialogue is possible because there is a sort of property in what is human that can be called *dialogicity* (Linell, 2009). This perspective expands the comprehension of what dialogue is beyond the interaction of two or more people verbally to consider, from within the person, the possibility of both heterodialogue: that is, dialogue with others, including imaginary others, and the dialogue with oneself, which may include even the dialogue between different I's (selves) that are constituted in a given time and in regard to a specific situation.

An "I" conceived as multiple dialogue of I's among which relationships of agreement or disagreement may exist, indifference or importance, competence or cooperation, correspond very schematically with the concept of *dialogical self*. Through dialogue, human beings build meaning and novelty not only about themselves but the world; they build meaning about what happens to them, what they know or about how they can act (Hermans, Kempen, & Van Loon, 1992).

From another perspective, also dialogical, Maturana states that conversing is a practice preserved ontogenetically as co-ontogeny. Conversing marks the properly human way of living, which implies the "languaging" intertwined with the "emotioning." Maturana states that this is part of the ontogeny phenotype defined the human being (Maturana, 2002).

Conversing is understood as an action pertaining to the emotional field in which language emerges as a way of being in coordination of actions in the intimacy of conviviality. What's human is always lived in conversation, and the wellbeing or suffering of humans depends on the kind of conversation it has. Maturana points out that all human endeavors happen in conversation, whichever the domain of the experience in which it takes place may be, from the one that constitutes physical space to the one that constitutes mystical space. Conversing implies consensual behavioral coordination; it's a consensual intertwining with an emotional flow that may also be consensual (Maturana, 2001).

The recursion may be understood in many ways, one of them is the one linked to language and how this can be recursive. To Maturana (2001, 2002), language is a biological phenomenon that implies behavioral coordination, in which participants coordinate their behavior in relation to their context and in regard of their own behavioral coordination. Only when there is recursion in the behavioral coordination, there is language. In human beings, for example, when somebody talks, another one answers regarding what the first has asked or affirmed and does so in many ways, with words or gestures. If he does not answer, there is not communication nor language as such. Thus, language creates a flow of coordination of actions, and that is the how it is known that a person is talking to another.

Recursion, to put it in a simple way, is a process that repeats until certain conditions are met. In the act of language, the process that repeats occurs when the first subject initiates a phrase in the conversation and the other one answers in relation or in coordination with what was said: What happens depends on the precedent. The recursion is maintained until the end of the conversation (Maturana, 2001). Recursion may be understood as a principle of functioning of behavior which can be discovered in the nature of social life. It is then a way through which life is sustained and reproduces.

The idea of recursion may be mostly understood if it is linked with other two equally essential ideas: that of auto-production and auto-organization, and the triangulation of these three principles: recursion, auto-production, and auto-organization allow us to understand complex systems, such as life, society, and human behavior, among other matters. Hofstadter (2007) states that human intelligence acts in a recursive way. Recursive systems have the characteristic of appealing to themselves—the same as human beings do when working with their thoughts to generate novelties about themselves (autopoiesis). Hofstadter (2007) states that recursive sequences have the intrinsic quality of assuming a behavior that is more complex each time, so that the most one advances, the predictability is less, as it seems to happen with human behavior; but at the same time, there is a higher opportunity for balance and for finding solutions that are outside parameters that threaten the system with its disappearance.

Upon leaving the mystic–ascetic reading and boarding a linguistic approach to the Ignatian proposal, new recursive aspects may be identified within it. Ignatius de Loyola, says Barthes, makes communication with God dependent on an inflexible order that builds this new language, a special language, founding for the divine interlocution. The new language, according to Barthes is built, depending strictly and rigorously upon (1) being isolated in order to find God, (2) articulating the five senses, (3) ordering, according to the retirement director's metric, and (4) theatricalizing to the point in which language becomes unlimited and recreates the christic tales as if they were medieval mysteries, scenes that invite the exerciser to live a scene from the written script (Barthes, 1997).

All these linguistic components would allow the subject not only to follow the speech contents but to be conscious of the emotions present in the internal dialogue experience and to be able to remain in it until uncovering what this experience may offer him. That is to say, according to Maturana (2001), the fundamental condition for recurrent interactions is to exist in the dialogical. Someone wants to dialogue and keep in the dialogue exchanging and answering other proposals if there is emotionality flowing from one point to another in the conversation, and that may grow as it becomes deeper. Emotions occur in the interactions upon experiencing what the other has to say and expressing what he means. The two emotions Maturana highlights as basic are those of rejection and love, which in the exercises arise at the moment of accepting or leaving the personal experience and has its maximum intention when making the choice. To Ignatius, the imagination is very relevant as an essential element of contemplations. In this respect, Barthes (1997) states that the mental image and discursive image proposed by Ignatius as nuclear part of the exercises becomes a linguistic unity. That is to say, the image becomes part of the unity of a language being constructed, and not anymore as a scale of a spiritual path where the subject melts in God's love, implying an exchange of a previous dialogue, but rather a dialogue with God.

In spiritual exercises, an experience of a dialogue with God is lived that may be judged repetitive by the type of order proposed and which was covered before. This repetition may be seen as recursive since what occurs and generates the subject is a consequence of its previous realization; in other words, it involves a process where the actions are linked—in some way coordinated, as happens in any dialogue where the individual answers based on what the other one states as a stimulus and vice versa, enlarging each time the environment of the relationship and interaction as the dialogue does not become exhausted with only one interaction. The exercises are proposed in a given order, and they should be done that way to produce the effect of interlocution with God. The way of articulating the experiences that allows the text to emerge off of each exerciser is the so-called Ignatian repetition. This way of coming back on each exercise is reproduced until the subject achieves change, generates novelty, and constructs meaning, as happens when an individual talks to another and the exchange stops when the parts have understood what the other wants to say or have opened a way to comprehend the other's point of view and arrive at something that may be given a sense of dialogue. For this reason, talking to God from the exercises is a recursive experience from a linguistic perspective.

The true change or potential transformation may or not occur at the summit moment, between the moment of deciding at the end of the second week and during the third week, which makes present the contemplation of Jesus' passion. This is the space that breaks with the linear structure and the three classic ways, creating an empty space that is neither questioned nor answered, based on the proposed dialogue and that, according to Barthes (1997), unchains a doubtful moment of indecision protecting the exerciser of the certainty of divinity. This is what is new in the way of interaction, making it dialogical and human. Dialogue, despite being structured and following a procedure that is in a certain way inflexible, does not necessarily arrive at the same point; it may create novelty at the end of the repetition. Ignatius identifies with this with Saint Augustine's glance in accordance to the search of the subject within himself, from his freedom. This process of relation proposed by the exercises marks two moments in the conversation and in the dialogue. The subject decides, and it can be recognized as recursion and not repetition (the third week seems to be a repetition of the first one) as it becomes from the previous situation and does not lead necessarily to the same place, repeating the cycle.

Recursion is also evident if one takes into consideration that the exerciser's interaction with God is a relationship of retroaction. Progress is made in dialogues initiating with a composition placed in a context, with an initial demand and which closes with intimate colloquies. Through questions and the imaginative process, the experience of the exercise enables finding elements to be used in a retroactive way. It is not defined at the beginning where the exercise will lead the person.

Lastly, the dialogical dimension of the exercises is presented, identified as coordination of coordination. The scheme shown in Figure 7.13 represents this dimension. In a determined context, God initiates the communication with the exerciser through X, represented by the interior arrow that

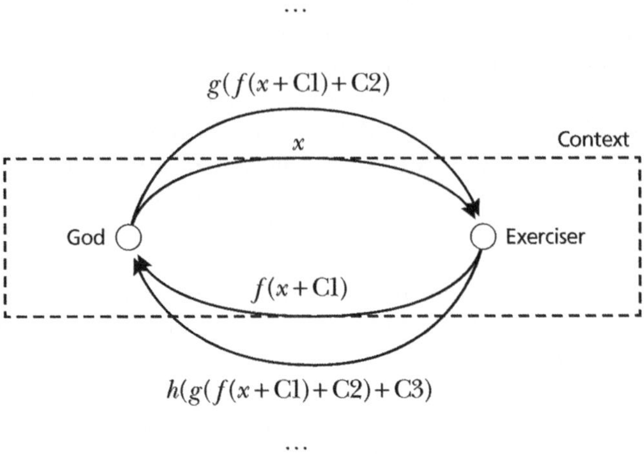

Figure 7.13 Dialogical structure of the exercises' objectives.

leads to the exerciser. The exerciser responds with $f(x + \text{C}1)$, in other words with a linguistic behavior depending on x and the context in which it is found, C1 represented with the internal arrow directed to God. Then, God manifests according to the exerciser's answer and the C2 context, which is the evolution of context C1 in which the previous interaction took place—in other words, in a recurrent way. This is shown with the external arrow lending to the exerciser and continues thus, successively.

The model allows the representation of eventual contextual variation among different behaviors that may correspond to changes in position of the exerciser during prayer, tiredness, and so on. It is necessary to clarify that functions are different each time, indicating the internal change the exerciser's conversation supposes. The ellipses (. . .) suggest further emissions represented with radial arrows that get bigger each time.

This schema can be identified on each of the previously analyzed levels. The different recurring structures, identified at a given hour of the exercise, establish this type of pattern. Particularly, from the method elements and the context, Figure 7.14 may be proposed.

According to Figure 7.14, God initiates the dialogue, being present in the exerciser's life, which will be considered the exercise referent. The exerciser, thanks to the guidelines supplied by the text of a contemplation or meditation, responds, experimenting with the consideration of the suggested points and projecting it (*reflectir*) in his or her life. In other words, the exerciser's answer depends upon the referent and furthermore on the environmental conditions under which the exercise is carried out.

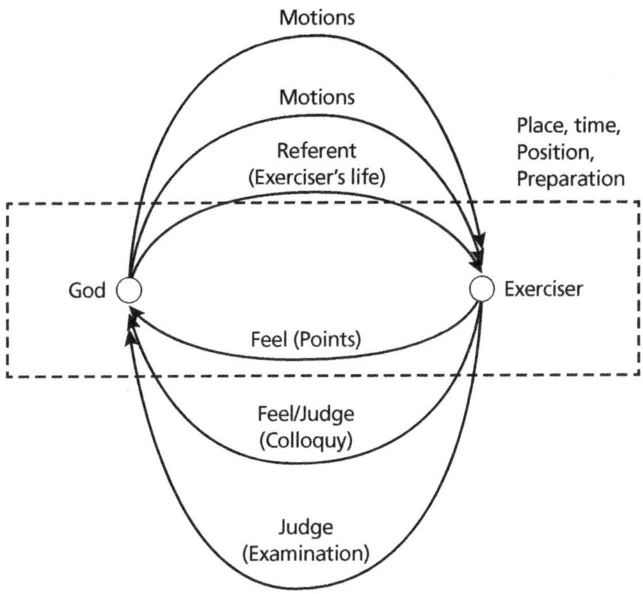

Figure 7.14 Dialogical structure of the exercises' timing.

As an answer, God generates motion within the exerciser that will exist in function of the projections the latter has made. After several iterations, the colloquy is carried out from what has been felt, which is a conversation that's carried out as it goes; that is, it depends upon the motion received previously. Here, the exerciser continues feeling and begins to discern the previous motions. Again, God responds by producing motions that will be in function of the path taken by the exerciser in the colloquy. Finally, upon ending the exercise, an examination is carried out in which one retraces everything that's been done. Then, the exercises' method and context produce a recurring structure of coordinated behaviors of coordination.

As a second example, the activity of the fourth day of the second week may be considered through the scheme in Figure 7.15, which shows that coordinated behaviors are behavioral coordination in two ways: because they depend on previous behavior of the day's conversational scheme and because they consist of the realization of the dialogical structure of the exercise represented in Figure 7.13.

Now then, in the same way dialogues are produced from recurrent elements that compose the Exercise's method and context, it is also produced from the order of them. For example, considering the full month the representation of Figure 7.16 appears.

The foregoing schemes do not exhaust all linguistic dynamics produced by virtue of other recurring elements of method and context in the weekly

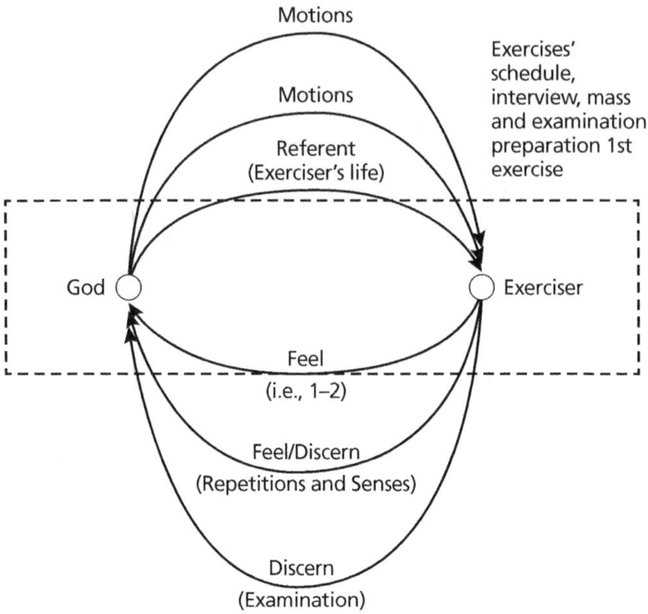

Figure 7.15 Dialogical structure of a day of exercises.

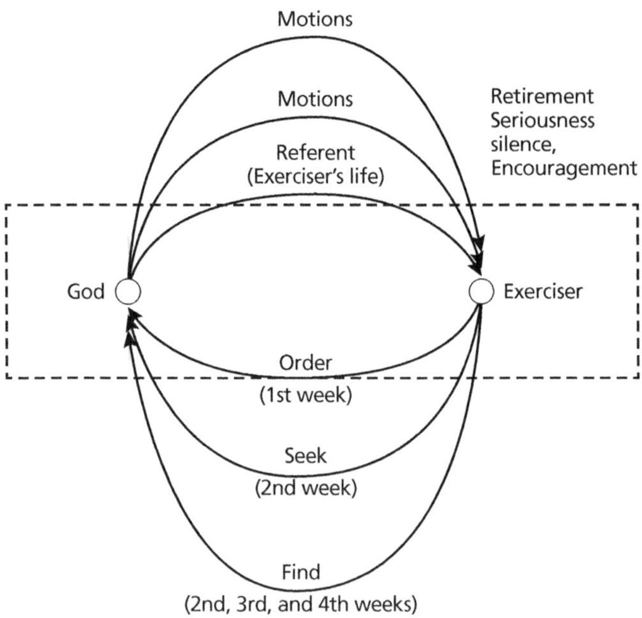

Figure 7.16 Dialogical structure of a month's exercises.

and monthly level, and by effect of the exercise level of the day and the week. Due to space considerations, it is not possible to include them in this work; however, it can be affirmed that there is evidence between the homothetic recursion and algorithmic of all the exercises that can also recognize the linguistic recursion arising from the dialogical character of the aforementioned.

CONCLUSION

The recursive structures presented come into relation, facing the fulfillment of the exercises' objective. Both the homothecy discovered in the exercises and its algorithmic recursion at different temporary levels are the conditions for the possibility of the presence of linguistic recursion. This is what allows *Spiritual Exercises* to surpass the isolated reading of the ascetic and the mystical, solving the tension they suppose. The result is an experience oriented towards the construction of the exerciser's identity.

A series of recursive operations are directed, thus allowing, according to Saint Ignatius' experience, the dialogue with God: a dialogue that forges the "I" and that in that sense builds the identity, an identity that is the encounter with what God wants for each one. The identity seen from the foregoing is the construction of the subject for the subject and in the case analyzed, comes from the illumination of God, who allows finding His will without imposing it.

Recursion in exercises facilitates the self-organization of the subject in the systemic sense, going back to what's been said quoting Hofstadter (2007). This is an experience of change generated in the limits of the human being immersed in his time. The exerciser is object and promoter of change outside the two pre-established places: that of the question to divinity and the pre-established answer. This place is that of the decision that occurs at the end of the second week and in the contemplation during the third week. This moment transforms and builds the identity in that the subject can make the ontological affirmation "the being that is" and can abstain or negate his identity affirming the being that is not, in front of the invitation of contemplating the passion and death of Jesus (Barthes, 1997).

From the developmental psychology standpoint, spiritual exercises may be considered as an example of a social practice, which through a recursive ordering of experience, and with a regular attendance explicitly to dialogue, pretends the creation of novelty in the path of personal development. Even though this practice is embedded inside of a religion and a particular historical context, recursive elements of the operation of culture through semiosis can be observed. The exerciser, the other, and God are part of a triad that dialogues in the search for a new meaning. Furthermore, the exercises show two key characteristics of psychical experience, underlined by cultural psychology: the first one, that the psyche occurs in the interdependence of the person and his environment with a significant social guidance, the psychological isn't something "internal" or isolated which produces results; the second, that the production of meaning is always confronted by the uncertainty of finding it or not. In this sense, the ordering the exercises produce, recursively, would be contributing to channel the production of meaning, but at no time guaranteeing it.

On the other hand, spiritual exercises may be considered as a sort of psychotherapy born before the invention of psychology, and in that sense it's comprehension, here as an initial step from a dialogical perspective and searching for evidence of recursion, that may be turned into a case study to establish comparisons between this religious practice and modern psychotherapy. Loredo and Blanco (2011), for instance, propose the auricular confession as a cultural practice, helping to understand how psychotherapy works today.

Finally, highlighting recursion in this social practice also serves to justify its interdisciplinary use. The application of analogies and metaphors on the properties for dynamical systems—among them, recursion, homothecy and fractality—widens the bases for finding the application of these and other concepts for a better interpretation of religious experience. This is the work to be done even in the still just-born field of cultural psychology of religion (Belzen, 2011).

NOTES

1. The quotes from the Spiritual Exercises of Saint Ignatious are taken from the version translated by father Elder Mullan, S.J. in 1914. The numbers correspond with the autograph, and are taken from Santiago Arzubialde, S.J. *Ejercicios Espirituales de S. Ignacio: Historia y Análisis.* Santander: Sal Terrae, 2009.

REFERENCES

Arzubialde, S., S.J. (2009). *Ejercicios Espirituales de S. Ignacio: Historia y Análisis.* Santander, Spain: Sal Terrae.

Barthes, R. (1997). *Sade, Fourier, Loyola.* Los Angeles, CA: Hill and Wang.

Belzen, J. A. (2011). La perspectiva cultural en la psicología de la religión: estudio sobre la espiritualidad "bevindelijke" a modo de ejemplo [The cultural perspective on the religion's psychology : study about the spirituality "benvindelijke" by exemplary]. *Estudios de Psicología, 32*(1), 103–130.

Hermans, H. J. M., Kempen, H. J. G., & Van Loon, R. J. P. (1992). The dialogical self: Beyond individualism and rationalism. *American Psychologist, 47,* 23–33.

Hofstadter, D. (2007). *Goedel, Escher, Bach: An eternal golden braid.* New York, NY: Basic Books

Linell, P. (2009). *Rethinking language, mind and world dialogically.* Charlotte, NC: Information Age.

Loredo, J., & Blanco, F. (2011). La práctica de la confesión y su génesis como tecnología psicológica. *Estudios de Psicologia, 32*(1).

Lop Sebastià, M. (2000). *Los directorios de ejercicios.* Santander, Spain: Sal Terrae.

Lora-Garcés, M. C. (2005). El goce místico y la escritura en una monja de la colonia [The mystic enjoyment and writing a nun of the colony]. *Poligramas, 22,* 21–40.

Lora-Garcés, M. C. (2005, June). Poligramas: Revista literaria. *El goce místico y la escritura en una monja de la colonia* (22). Cali, Valle, Colombia. Obtenido de El goce místico y la escritura en una monja de la colonia: www.poligramas.univalle.edu.co

Loredo, J. C., & Blanco, F. (2011). La práctica de la confesión y su génesis como tenología psicológica [The practice of confession and its genesis as a psychological technology]. *Estudios de Psicología, 32*(1), 85–102.

Mandelbrot, B. (1987). *Los Objetos fractales: forma, azar y dimensión.* Barcelona, Spain: Tusquets Editores S.A.

Maturana, H. (2001). *Emociones y lenguaje en educación y política.* Santiago, Chile: Editorial Dolmen, Ensayo.

Maturana, H. (2002). *Transformación en la convivencia.* Barcelona, Spain: Océano.

Melloni, J. S. J. (2001). *La mistagogía de los ejercicios.* Bilbao, España: Mensajero Sal Terrae.

Mullan, E., S.J. (1914). *The spiritual exercises of St. Ignatius of Loyola.* New York, NY: P. J. Kenedy & Sons.

Order of Friars Minor. (1909). In The Catholic Encyclopedia. New York, NY: Robert Appleton Company. Retrieved from http://www.newadvent.org/cathen/06281x.htm

Sierpinski, W. (1974). *Oeuvres choises.* Varsovie, France: Éditions Scientifiques.

Valsiner, J. (2007). *Culture in minds on societies.* Thousand Oaks, CA: Sage.

CHAPTER 8

MARKING THE PAST FOR THE FUTURE

Roadside Shrines and Recursivity

Zachary Beckstead
Grand Valley State University

In the United States, driving and the road evoke mythical and epic images of highways along the coasts, Route 66, American towns with antiquated churches and downtowns, road trips, and mountains and rolling hills. Travel by car has been associated with a quest for freedom, power, and exploration of the landscape and self. Yet the reality of most everyday travel is that driving in a car is a mundane event in which we move from home to work, work to the grocery store, and back home. We might sandwich some trips in between our ordinary mode of traveling, but rarely do we pass by scenes and objects that captivate our attention. Instead, in our modern landscapes we primarily encounter gridlock, strip malls, and billboards. One exception to the monotony of the road is the roadside memorial. As I drive or ride as a passenger and encounter the occasional roadside memorial, typically with a cross and other periphery decorations, my heart momentarily sinks and my attention is briefly arrested. Who died there, and how did they die? Who

Cultural Psychology of Recursive Processes, pages 219–239

placed the memorial, and do they visit this site often? My response is usually fleeting and more affective than cognitive—feeling of a sense of dread and apprehension. I then pass by and the flicker of feeling falls into the background. Still, I wonder why I am fascinated and troubled by these objects.

This chapter offers an exploration on this particular act of commemoration: roadside memorials and shrines. In particular, I am interested in investigating how these frequently simple and occasionally elaborate objects captivate and arrest our attention and what they communicate to those who create, maintain, and pass by these material artifacts. Of course many signs, objects, and practices relating to death are sure to be experienced as disturbing and fascinating. Thus, what this chapter explores is how these memorials can pack an emotional punch in spite of (or perhaps because of) existing at the margins and even though they are mostly encountered quite fleetingly. What I will suggest is that the roadside memorial and cross create connections between the past, present, and future and link the living and the dead through dynamic and recursive processes. It is typical to say that roadside memorials and crosses make grief manifest; more to the point, I argue, they make temporal and social relationships explicit and provide the context and ground for our emotional engagement and perspective taking. Because of the roadside memorial and its capacity to evoke and unite different perspectives and positions, human beings can enter into the imaginative realm and feel and contemplate the perspectives of the bereaved, the dead, and the witnesses of these objects.

In the first section of this chapter, I will examine the memorials and the landscape in general and then roadside memorials and crosses in particular. Next, I will demonstrate the recursive aspects of the practice of placing these markers and the meaning-making activities that they trigger. In order to accomplish this task, I will link David Herbst's co-genetic logic with notions of recursion. Ultimately, I conclude that recurisivity and recursive processes are a social and relational phenomenon that both allow us to distance ourselves from the immediate environment and feel ourselves more intimately related to the environment.

MEMORIAL LANDSCAPES

Our environments bear the marks of the past—traces of tragedies and triumphs, heroic acts and demonic plots realized. While these "scars" heal as people rebuild and rehabilitate the damaged land, quite often memorials or shrines are left behind to keep the memory of these events in the public's consciousness. Furthermore, events that occurred in distant lands are marked and remembered in war memorials such as the Vietnam Veteran's Memorial in Washington, DC. In any case, it is through these various memorials that the environment is transformed and set apart as a sacred or quasi-sacred place (Foote, 2007) that, for some, is deeply moving and affecting. Moreover, these places are seen as key to the grieving process by allowing individuals or groups to externalize and make manifest their powerful emotional responses, receive support, and adjust to life without the deceased. In the case of 9/11 or the Vietnam Veterans' Memorial, these memorials also are believed to help the public deal with socially traumatic events caused by unexpected losses and/or controversial events. While the forms of commemoration and events remembered vary greatly, one of the universal aims is to keep the deceased and the historical event in the public consciousness, to keep the memory alive (Winter, 2006). This desire can be maintained for different reasons and utilized by individuals and institutions for different ends. For instance, it is not uncommon for social and civic institutions to utilize memorials and monuments to guide individuals and their citizens toward certain ways of thinking, feeling, and acting (Beckstead, 2012). Memorials and monuments are instruments or tools in the development of actions and sensibilities that are connected with didactic, therapeutic, or propaganda purposes (Beckstead, Twose, Levesque-Gottlieb, & Rizzo, 2011). Crucial to this process is the material markers that encode the memory of the deceased and past events.

While it is understandable that much attention has been given to grand memorials and monuments, smaller shrines are equally interesting and significant in human lives. One example of a smaller shrine or memorial practice is the roadside cross that marks vehicular accidents and unanticipated loss. As tangible markers of loss, roadside crosses are simultaneously reminders of a life (or lives) cut short, the grief of those left behind, warnings of the dangers of the road, and the suddenness and inevitability of death.

ROADSIDE MEMORIALS AND CROSSES

Roadside memorials are becoming a common sight in both rural and urban landscapes in the United States (Everett, 2002). While most roadside shrines and crosses are memorials to those who died in an automobile accident, others commemorate those who died by other forms of accidents, natural disasters, and homicide. On a well-traveled road near my house in Michigan stand three roadside crosses next to each other (see photo below). Houses, a daycare center, small businesses, and a path that leads to a nearby river all line the road near the crosses. One memorial is made up of two smaller crosses—one painted white and other brown—that have been placed in front of a larger wooden cross. The larger cross has the name "Bobby" written across the horizontal board and "R.I.P" and "Son" across the vertical board. Flowers sit at the base of the crosses. Roadside crosses and shrines vary in how elaborate they are—some simply include a cross while others are ornate semiotic complexes. Many crosses will include the date of birth and death of the loved one and others include pictures, objects representing activities enjoyed by the individual, flowers, photographs of the deceased, teddy bears (for especially young victims), and personal notes. Roadside crosses are a cultural phenomenon, and therefore a commemorative template has emerged rooted in conventional symbolic systems; however, as they mark the loss of an individual, they are particularized and often quite idiosyncratic.

As a cultural practice, roadside assemblages entail general social practices and material patterns that cut across social groups and boundaries. Roadside crosses can be found in Australia, Canada, South and Central America, and other countries. According to many accounts, the practice of placing a cross at the scene of an accident in the United States context can be traced back to the burial practices in the Southwestern part of the United States and connected with Native American and Mexican American peoples. As the funeral party carried the deceased to their final resting place, they would stop and recite prayers or the rosary along the way. The locations of these brief pauses were marked by crosses or *descansos* ("*rest*" in Spanish) as they were called (Everett, 2002). However, the practice is not limited to Native Americans, Latinos, or Catholics and, as mentioned above, is found in many countries (Owens, 2006). In Chile, there is a belief

Photo 8.1 Roadside memorial in Walker, MI.

that the spirits of accident victims remain at the site since they are troubled. Roadside crosses therefore allow family members, friends, and community members to pray and ease the suffering of the individual. Maida Owens has demonstrated how roadside crosses have become more frequent as reports and stories in the media proliferated over the last few decades. It seems these stories have sparked interest in the practice and ultimately have provided a cultural template mourning the dead. In more recent times, crosses reemerged in parts of North America as a form of memorial on the road through the organization Mothers Against Drunk Driving (MADD) to bring awareness to this social problem.

ROADSIDE CROSSES AS SPONTANEOUS SHRINES

Interest in public memorials and shrines has blossomed recently in both the general public and academic circles, especially in folklore studies and anthropology. One reason for the reemergence of this interest and awareness can be traced to a series of very visible tragic events during the last twenty years. Diana, Princess of Wales' death in a car accident generated an outpouring of grief that was made manifest by the number of flowers, candles, and other objects placed on the *Cours la Reine* in Paris. Similarly, one of the spontaneous responses to the bombing of the Alfred P. Murrah Federal Building in Oklahoma City by thousands of individuals was the placement of personal mementos—flowers, notes, pictures, toys, teddy bears, and other items—on the chain link fence that was initially set up to cordon off the area. This fence, which was meant to be temporary, still exists today with new items being frequently added. Additionally, the events and destruction of 9/11 were marked, almost immediately, by a vast array of shines and memorials in New York City, Washington DC, and Pennsylvania. More permanent memorials have been constructed at these sites; however, they have not replaced the more informal and personal commemorative contributions made by individuals.

It is because of the characteristics of these memorials—personal, informal, and not planned or erected by governmental institutions—that Jack Santino describes them as *spontaneous shrines* (2006). In contrast to top-down, state-sponsored memorials, these are created by ordinary citizens, lack state sponsorship, and emerge quickly in public setting. These shrines therefore seem to indicate "genuine" outpouring of grief that emerges without institutional support or coordination. Even at a state-sponsored memorial such as the Vietnam Veteran's Memorial in Washington, DC, we can observe mementos and gifts left at the Wall of Names. Flowers, notes, poems, uniforms, and small objects like teddy bears are left behind every day and are collected by the staff (Wagner-Pacifici & Schwartz, 1991).

The emergence of spontaneous shrines for Princess Diana, the wall at the Oklahoma City bombing memorial, and countless roadside crosses demonstrate that individuals with varying degrees of relatedness participate in this process of commemoration. Moreover, these shrines have an *outlaw* quality to them; often in the case of roadside shrines and crosses, they are often prohibited or at least highly regulated institutions, ostensibly for safety and road maintenance concerns. Roadside crosses have the potential to distract drivers, or so the argument goes. While these shrines emerge "spontaneously," they occasionally are co-opted by institutions in order to assist in the agendas these institutions promote. For instance, roadside markers can be purchased from the Department of Transportation in some states in the U.S. Additionally, placement of roadside crosses has been challenged in the United States because critics argue that they violate the separation of church and state and constitute the government promoting a particular religion (Christianity) since the crosses are usually on public land. Thus, roadside crosses and shrines are objects of public scrutiny and contestation.

Memorials, Shrines, and Material Culture

Roadside shrines and crosses are closely related to larger memorials such as the Vietnam Veteran's Memorial or the Oklahoma City Bombing Memorial in becoming sites of pilgrimage with personal and vernacular items left behind (Santino, 2006). Radical and often unanticipated loss is marked and the unknown, uncertainty, and fragility of life are confronted through these memorials and our encounters with them. These memorials mark ruptures of everyday living and thus contrast with the everyday landscapes of our urban and suburban landscapes. According to Yi-Fu Tuan (as cited in Everett, 2002, p. 12–13):

> Contemporary space, however colorful and varied, lacks polarized tension between the numinous and quotidian. Contemporary life, however pleasant and exciting, moves on one plane—the plane encompassed by rational and humanist vision. Ecstasy and dread, the heights and depths, the awesome and the transcendent rarely intrude on our lies and on our landscapes except under the influence of chemical stimulus.

Tuan's reading of contemporary space is similar to Max Weber's discourse on *disenchantment* in modernity. According to Weber (1922/1993), the rise of secularism has coincided in the decline of viewing the world from a religious frame that saw the world as barely veiling a numinous, spiritual, and enchanted realm. Thus, roadside shrines can be understood as setting apart the mundane, contemporary space of the road from the sacred or hallowed (note distinction) space that is lacking in the modern day. While the road is a liminal phenomenon (*between and betwixt* home and our destination) (Beckstead, 2012; Turner & Turner, 1977; Valsiner, 2007), the site of a vehicular accident resulting in death becomes a threshold between this world and the world of the dead. The cross marks and engenders a rupture where the ordinary and extraordinary and the mundane and divine meet. Roadside shrines, therefore, construct a place where communication, commemoration, and reflection beyond the "ordinary" or "mundane" can occur (Everett, 2002). In this sense, they function to elicit experiences that transcend ordinary life.

MULTIVOCALITY OF ROADSIDE CROSSES AND SHRINES

The meanings of these shrines and memorials are polyvalent and depend on the situation under which the accident occurred (i.e., drunk driving vs. falling asleep at the wheel), who establishes and maintains the memorial (i.e., family vs. friends of the deceased), and who encounters the memorial (i.e., "ordinary" citizens vs. governmental agencies) (see Owens, 2006, for a discussion). Furthermore, while the roadside cross may convey certain messages based on the religious and cultural background of the members of a particular society or group, the meaning or meanings of the roadside cross and death site depend, in large part, on the many potential perspectives/positions and participants who create, encounter, and manipulate (in one way or another) the material assemblage of the death site. Indeed, just as the cross is the symbolic and material anchor of the site, the roadside cross and shrine constitute a powerful and dynamic symbolic coordinate for the friends and family of the deceased, the passing drivers, the wider community, and also government institutions (e.g., departments of transportation) to coordinate their activities, negotiate and create meaning, and become affective guidance mechanisms.

> So, too, roadside memorial markers offer a meeting place for communication, remembrance and reflection, separate from the "everyday." Embracing many voices, they may also represent the quiet acquiescence of civil authority, for in many states their mere presence violates official policy. The multivocality and cooperation embodied in each memorial, and the vernacular support that facilitates their existence, contributes to their dynamism and popularity. (Everett, 2002, p. 14)

Roadside memorials bring to the foreground reflections, images, and expressions of death and grief, and their creation and reproduction involve notions and social representations of how human beings should remember the dead and make sense of their passing. One common assertion made about roadside memorials and similar practices is that they reflect public expressions of private grief. While this is certainly true to an extent, we could argue that this assentation highlights, isolates, and emphasizes the *individual* who has died. In essence, this practice reflects modernity's relationship and attitude toward death—individualism is ascendant in both life and death (Aries, 1974). Other scholars make this argument in reference to larger memorials. Many scholars argued, regarding the Vietnam Veterans Memorial in Washington, DC, that the significance and power of the memorial will quickly fade since the focus of the memorial is on the individual soldiers and not on the conflict and social issues (see Wagner-Pacifici & Schwartz, 1991, for a discussion of this issue).

In spite of these proclamations, some scholars suggest that memorials are social processes that open up dialogue between different parties or stakeholders (Santino, 2006), yet the truly dialogical nature of memorials is still often in the background and underdeveloped. In the following section I want to challenge the notion that roadside memorials are primarily "public expressions of private grief" and make salient the public, performative, and recursive nature of these memorials. One of the powerful and most salient characteristics of what are now called *spontaneous memorials or shrines* (Santino, 2006) is their ability to make apparent more than private emotions but also social relationships and perspectives. This recursive capacity generates or opens up the possibility of dialogue, perspective-taking, and empathy. This generative capacity can be conceptualized as a recursive process that creates a system of relationships and links the parts together and to the whole. Ultimately, the roadside cross opens up a web of relationships and positions that individuals can adopt and negotiate. Below I want to flesh out the recursive characteristics of these memorials and suggest that roadside crosses and similar memorials promote dialogue through their recursive nature.

RELIGIOUS AND NON-RELIGIOUS SIGNIFICANCE OF ROADSIDE CROSSES

Roadside crosses are self-evidently religious symbols and markers connected to the Christian tradition. Although other material forms of materials are used in roadside memorials and shrines, they overwhelmingly include a cross in their assemblage. Crosses are religious symbols rooted in the Christian tradition. Thus, for believers, they are related to the life, death, and resurrection of Jesus Christ. Just as all men and women die, it is through the suffering and death of Christ that all men and women may be redeemed and resurrected. Notions of death and rebirth are central to the cross, yet the cross is not widely accepted by some Christian denominations. Some protestants find the cross is too closely related to Catholicism and idolatry, and the Church of Jesus Christ (Mormons) eschews the cross as a major symbol because of its focus on the death and suffering of Jesus. Still, the cross has come to stand for hope, rebirth, and even simply a widely recognized marker that a death occurred at a particular place. Because of the latter reason, even non-Christians or non-practicing Christians tend to erect a cross at the site of a fatal accident (Everett, 2002).

RECURSION, RECURSIVITY, AND RECURSIVE PROCESSES: TYPES AND CHARACTERISTICS

Notions of recursion and recursivity provide a potentially useful conceptual framework for analyzing the complexity and dynamic person < > environment and the transformative potential of the roadside markers. As discussed in the introduction and throughout these chapters, recursion, recursivity, iteration, and similar concepts can be found in mathematics, linguistics, and some areas of the humanities, and they have been recently explored in other branches of psychology. Here I want to clarify how I am using the term as I explore the recursive aspects of spontaneous shrines such as roadside crosses and our encounters with them.

Recursivity is etymologically related to the idea of *return* or *returning*. As Adolfo Perinat illustrates (see Chapter 2), the Latin word that recursivity stems from, *recurree*, implies that one returns to the beginning by retracing the same path. The path is not linear, but rather it is a loop; hence the notion of recursivity involves the notion of the self looping back on itself and thus creating itself (autopoesis). Here we see that the recursive act of looping is a creative process; when one returns to the beginning, a change or transformation occurs (i.e., self-awareness is the product). Self-reference and circularity have been the defining characteristics of recursion in the humanities and social sciences. Steven Pinker and Ray Jackendoff (2005) define recursion as a "procedure that calls itself, or... a constituent that contains a constituent of the same kind" (p. 203). According to Clifford, Friesen, and Jardine (2001), one prominent way to way that recursion has been conceptualized in the humanities is the mind's circular, and self-reflective capacity: Recursion "refers to the act of a mind or self 'looping back,' 'turning around' or reflecting on itself, and in this way actually creating itself as a conscious self—the highest expression of human awareness" (Clifford et al., 2001, p. 2).

This understanding of recursivity highlights the reflective and metacritical faculties of human beings. We not only can have thoughts about thinking, but this "looping" capacity allows for the self to distance itself from the flow of the immediate here-and-now experience and setting. Most importantly for the purposes of this chapter, recursivity points to the ways through which human beings relate to their environment. Thus, recursion is a process through which self-referencing, embedding structures (e.g., in language), looping, circularity, and transformation occur. Recursion thus is best seen as a process generating change and novel forms. Things, whether they are words, numbers, or other material agents, are not simply brought together and aggregated or repeated; rather the circularity produces non-linear or nonadditive changes. Again, as Chomsky (1957) noted within the context of language, recursion is generative in allowing for new and novel forms to emerge based on previous forms.

Here I want to emphasize an alternative, if not complementary, view of recursion: "Thought of this way, recursion is not a reflective turn of mind, but the thing's own variety. It is not organized around our volition, but rests in the character of things and how they constitute the world. They arrive, they make claims beyond our willing and doing" (Clifford et al., 2001, p. 5).

Drawing on hermeneutics, Clifford et al. (2001) argue that things themselves have recursive characteristics and suggest that recursion is not simply an act of the mind looping back on itself, but rather involves things looping back on us and "laying claims to our attention" (p. 3). Hence, I want to situate recursive processes as acts involving human agents, yet not reducible to the human being (i.e., it is more than a property of the human being). The implication is that recursion or recursive processes can be understand as a form of dialogue, or perhaps the basis for dialogue, between people and also between people and their environment. Moreover, this dialogue can be triggered by either agents, the person, or an aspect of the environment. Situating recursion or recursive processes in these terms would imbue them with a social meaning that has previously been overlooked. Most importantly, the notion of recursion or recursive processes emphasis the dynamic or generative aspect of these phenomena; that is, new forms emerge through these recursive acts and elements. Novelty, therefore, is a product of recursive acts and transformations.

Herbst's Co-Genetic Logic

In order to flesh out the recursive characteristics of roadside crosses and similar markers, and to illustrate how objects such as roadside crosses are complex and polyvalent, it will be helpful to briefly examine David Herbst's (1995) co-genetic logic. Herbst postulated that any time a distinction (differentiation) is made, three properties are co-created. We see when a circle is drawn (see Figure 8.1) the simultaneous emergence of three elements: the inside, outside and boundary. Previous to the drawing of the circle there was a blank space. But as we make the distinction, the inside is differentiated from the outside and the boundary emerges to unite them. Herbst

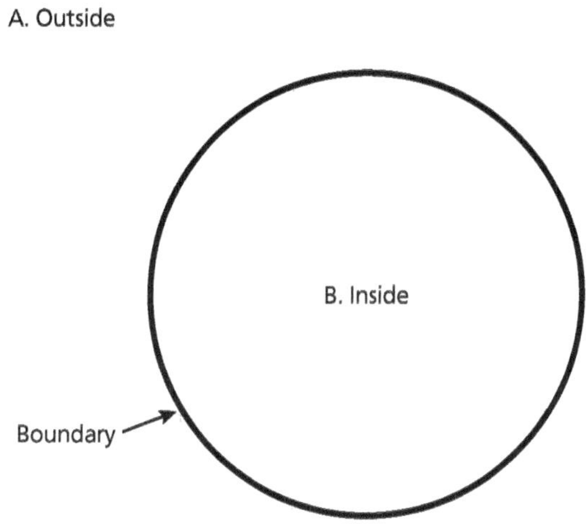

Figure 8.1 Herbst's primary distinction.

elaborates four principles of the primary distinction. First, he notes that it is co-genetic when the parts come into being together. Second, it is nonseparable since the parts cannot be taken apart and it cannot be constructed by bringing the parts together. Third, it is nonreducible because if one element is removed, all elements disappear. Fourth, the primary distinction is contextual. None of the parts or elements has an intrinsic essence or identity; rather, they are defined in terms of each other. The inside, for instance, has its characteristic only because of its relationship to the boundary and the outside. Based on these implications, Herbst resists traditional modes of analysis that break wholes into the components and elements and understands them as separate entities. As Herbst asks, "Or, if we have a wife and a husband, do we first and independently have a wife and a husband, and then link them together by marriage?" (1995, p. 69). Thus, Herbst conceives of phenomena being comprised of nodes of a network, and his logic is contextual and holistic.

Yair Neuman (2003) extends Herbst's co-genetic logic by suggesting that the primary distinction constitutes a recursive-hierarchical structure. Adapting Bateson's notion of *recursive-hierarchy*, he argues that the primary distinction as a complex, integrated whole that is composed of three subordinate units. The meaning of the whole is determined by the parts that operate together. Additionally, the complex whole is not static but recursive since each part is defined in relation to the other parts as noted above (self-referentiality). Neuman notes, "Therefore, we are not speaking about simple circularity that exists on a single level of analysis, but about a recursive-hierarchical structure in which the whole generates the meaning of its sub-units and the sub-units generate the meaning of the whole in a continuous process" (2003, p. 101).

The meaning of a subpart becomes possible only when the context is taken into account. The interpretation of a sign is accomplished by tacking back and forth between the whole and the subparts. For instance, understanding what "the cat" means depends on not only the dictionary meaning of a word but also the context (whole) of its use. Neuman notes that a cat could be an animal, the name given to a burglar, or a new punk band. This notion highlights that meaning is not fixed but instead is dynamically related and determined by the whole–part relationship. This view is significant because one implication is that what the roadside cross or memorial *means* is dependent on the relationship (or better, relating) of the person to the memorial (elaborated below). Significantly, for our purpose, this relationship is constructed by or grounded in the very act of placing a cross or material assemblage. Crosses and other material markers thus are, or function as, *boundary objects* that create new temporal configurations and social relationships in a recursive fashion.

Creating Connections—Past, Present, and Future

We can also look at the minimal roadside cross that is simply a cross on the side of the road. What emerges in the placement of this marker is a complex web of part–whole relationships that follow Herbst's co-genetic logic and these relationships are recursive. On the one hand, placing a marker at the site of an accident creates a distinction between the present, past, and future. The cross is a marker that creates and links the past (*what happened*) and the future (*what may happen*). In other terms, with roadside shrines, the past is "inserted" or grafted into the present through indexical, iconic, and symbolic signs. For instance, it is rather common to preserve the "scars" (indices) of the accident. Likewise, some roadside assemblages also contain the remnants of the car—a hubcap, broken mirror, fender, or other part of the car. Often photos of the deceased (iconic signs) and birth and death dates (symbolic signs) accompany these roadside memorials and serve as signs of the dead, loss, and a tragic event from the past. However, the cross with its religious symbolism offers hope and gestures toward the future when the deceased will be reunited with their loved ones. Furthermore, material markers transform in the present, loop back, and transform the past by renegotiating the past meanings of the accident. Thus, the material markers loop booth forward and backward.

What allows us to understand roadside shrines and crosses as recursive phenomena can be illustrated by comparing the creation of roadside shrines and to litter or billboards and traffic signs and examining how each transforms the environment. In the latter case of litter, objects intentionally tossed out of a passing car or by a pedestrian (or unintentionally flying out of the bed of a truck) alter the landscape creating a sense of "ugliness" or "blight" in the minds (thoughts and affects) of those who drive by. Litter is a sort of pollution and also may be a violation of laws in same areas and punished with fines. While litter can be seen as an indexical sign (human beings were passing by here and have left their mark), it does not gesture to a previous event in the same way a roadside cross or marker does. Litter is transformational but it is not self-referential toward the past in the sense of a memorial. In both instances, the present location and setting is reconfigured through the actions of human beings; however, litter does not function to trigger the looping process in evoking past events (e.g., car accidents and fatalities), the future (e.g., mourning acts) and relationships (e.g., deceased < > family and friends). Litter is a transformational addition to the environment, but a memorial is both an addition and subtraction as a memorial marks the absence or loss of life that *occurred at that particular spot.*

Similar to litter, billboards and other traffic signs line many of our roadside environments. In the case of the billboards, communicative signs are intended for the passers-by and attempt to canalize their thoughts and actions. With billboards, within the liminal space of the road, the billboards are embedded and attempt to guide the future actions of those passing by. Thus, billboards and litter are obviously different kind of material and symbolic objects. A billboard advertising for McDonald's draws on indexical, iconic, and symbolic signs to persuade drivers passing by to visit their restaurant. These billboards are located at the periphery of the view of those in the vehicle, but their redundancy seeks to capture the affect and volition of those encountering the sign. In comparison to roadside crosses and memorials, the billboard is predominately future-oriented (even if it advertises historical events) and is not placed at the particular location because of what had happened *at that particular site* before. As with litter on the side of the road, the past nor the future do not reverberate as they do with roadside shrines because the setting of the billboard is irrelevant. It is embedded in the environment, but it does not transform the setting the same way a roadside shrine does. Thus, the roadside cross creates the possibility of *reflecting* on the past, present, and future.

CREATING CONNECTIONS: PERSPECTIVE TAKING

Additionally, the roadside cross or assemblage creates and establishes a novel set of relationships between individuals and groups and social positions and perspectives. In Figure 8.2, we see the possible (recursive) relationships between the living and the dead that emerge through the placement of a

Living (A) and Dead (B)
Family and Friends of Deceased (A1)
Passers-By and wider Public (A2)
> Indicates directionality of relationship (i.e., B > A would illustrate how the living are relating to the dead)

B > A
- B > A(1)
  - A(2) > B > A(1)
  - A(1) > B > A(1)
- B > A(2)
  - A(1) > B > A(2)

Figure 8.2

roadside cross. On the left side we see the actor that is in the foreground (dominant) and on right side is the actor taking the perspective.

B > A: Obviously the reverse, A > B is not possible to analyze: the dead cannot yet be contacted to ascertain their perspective on the living. Instead, what is available for analysis and observation is the reflections the dead communicated through the signs of the shrines and ritual practices related to them. For instance, we may focus on the life that was tragically cut short, possibilities that will never be actualized, and relationships that have been dramatically altered (and some may see as severed).

B > A(1): Beyond the general notion that roadside crosses establish or establish the positions of both "living" and the "dead," we can further differentiate these positions based on the nature of the relationship to the deceased. Here we use A(1) to denote individuals intimately familiar with and related to the deceased. These refer to people who establish the roadside markers or who are closely involved in their planning and emergence. For these individuals, the focus of the memorial is foremost on the deceased. It reflects personal loss and grief and through icons (i.e., photos), symbols (i.e., teddy bears), and indices (i.e., material remains of the vehicle) that recall memories of a loved one's life and death. For these individuals, the markers function to connect them with the deceased in powerful and intimate ways.

B > A(2): Of course not all of those who encounter the roadside cross are intimately familiar with the deceased. Furthermore, most individuals who pass by these markers do not know whom they commemorate. However, the roadside cross as a boundary object establishes them as the *living* in obvious contrast to the *deceased*. Prior to encountering a roadside shrine, one is simply walking or driving; yet, noticing the roadside cross confers a new albeit temporary position on the passerby as someone who is not dead but who is very much alive (and taken-for-granted). Thus, for these individuals, roadside crosses might evoke a general sense of death, of one's own future death, and often a curiosity for what happened to cause the accident. Again, one might focus on the deceased, on the earthly future that has been extinguished and relationships severed, or the reflection might loop back and one may consider one's own mortality. This relationship and perspective is general and abstract; general ideas regarding the deceased belong to this level. The deceased, therefore, convey ideas of more general notions of death and mortality.

A(1) > B > A(2): Individuals passing by might not only focus on the deceased, but also on the living as well. One might turn their thoughts to the living that have been "left behind." Even if one does not know the circumstances of the death or the friends or relatives who erected the cross, one knows that the dead left loved ones behind and they cared enough to place the cross on the side of the road, to elaborate with material and symbolic objects, and to ritually renew the memorial by making updates (i.e., replacing flowers). Here we see that, as Santino notes, spontaneous shrines connect the living with the dead, especially with those who have been left behind.

> Spontaneous shrines both construct the relationship between the deceased and those who leave notes and memorabilia, and present that relationship to visitors. This is manifested in the notes and in the nature of the gifts which are brought, left and publically displayed.... The gifts have personal meaning, and this is indicative of—that is, they index—the nature of the relationship, real or (as with Princess Diana and other celebrities) imagined. Imagined, but no less felt. (Santino, 2006, p. 13)

What is established, therefore, is not something deeply interior, but inherently personal and relational. A cross, shrine of teddy-bears and other toys, photos, and other mementos are tangible links (observable to others) of the relationship between those who mourn, their dead, and those who observe these objects and acts of commemoration.

A(2) > B > A(1): Roadside crosses are constructed for the dead, but they are placed in very accessible and public settings. In contrast to cemeteries and other locations dedicated to the memory of the dead, the cross will be viewed by many other community members and non-acquaintances. This is significant since this form of communication now incorporates a completely different range of individuals, A(2), and the wider public with differentiated social roles and institutions. Those constructing roadside crosses may keep in mind and take the perspective of government institutions who may object to their existence and "random strangers" who have no connection or information of the life and death of their loved one. We can then suggest that personal public memorials and rituals gain their value by the (real and imagined) reactions that they create for others; the efficacy of these memorials for those in mourning lies in part in witnessing or imagining how others respond to them. Thus, one *cares* by demonstrating concern that is recognized as *care* through public and culturally shared symbols and actions. Here one may take the perspective of the other on him- or herself and imagine how others view him or her (see Gillespie, 2006; also Long et al., this volume, Chapter 4).

A(1) > B > A(1): Similar to the configuration above, those mourning a loved one may also communicate to others close to the deceased their concern, care, and attention for the dead. They may want to other family members to know that they love and miss their son or daughter, for instance. One example that we can cite is that of a case in Westgaard (2006) where a young adolescent was struck and killed by a bus outside of his school. His friends erected and maintained the cross as they said, "because you want it to show [that someone cared and mourned their friend]" (p. 151). In particular, they wanted to make it evident to their friend's parents that their son had friends who were fond of him and mourned his death; hence their friend would not be forgotten.

PRELIMINARY EMPIRICAL SUPPORT

Roadside shrines and crosses are complex and give rise to a multiplicity of meanings. Nevertheless, in order to make an initial exploration of what roadside crosses communicate, I asked students at a small, liberal arts college in the Southwest to respond to a few questions about roadside crosses and shrines. Since no participant mentioned having erected a shrine or cross for a relative or friend, we our observations are limited to the general A(2) position. Many participants commented on the connection between the roadside cross and the site of death.

> I typically always think about how the accident occurred and who was involved.
> Someone must have died or was badly wounded there.
> That someone died in a car accident near the cross.
> I hope it wasn't a drunk driver who killed an innocent child.

Additionally, and based on the analysis above, we would expect participants to have abstract, general interpretations of these objects (e.g., they relate to death) in terms of the B > A(2) relationship noted above. However, we would also expect participants to make sense of these objects as manifesting the relationship between those who created them and those died (noted above as the A(1) B > A(2) relationship. Table 8.1 presents some representative statements from participants, categorized by general utterances about death or their feelings about it (e.g., sadness) and also utterances about the living who put up these memorials.

Not surprisingly, many participants commented on how encountering roadside crosses and shrines evokes sadness related to the loss of life that the cross marked. Roadside markers unambiguously mark a place where someone lost his or her life and leads one to wonder or imagine what happened (i.e., was it a drunk driver who killed a child) and to reflect on large, overgeneralized notions of God, heaven, and death. However, in addition to these large meanings, roadside crosses index the relationship between those who created the markers and those who died. The roadside cross is a sign of death, but also that someone is remembering that individual, has not moved on, how much pain someone might be experiencing, and how much they care about their loved one or friend.

TABLE 8.1 Representative Statements

B > A(2)	A(1) > B > A(2)
Death, remembrance	That someone died on that spot, and that someone loved and missed them enough to set it up
Someone's death involving a car	A fatal wreck, someone showing remembrance to a loved one
A sad car wreck ended up with someone losing their life	When I see them I think about who may have passed away and I take a moment to pray for them and their family
Danger and sadness	Someone who had a wreck and died near that spot. Their friends and family trying to show where there spirit is
I think drunk drivers must have done something stupid	That someone died there and that their family hasn't moved on yet
Death, God, heaven	That someone is remembering a person or event
Death, sadness	Who/how did someone die there, and how sad that must have been for family and friends
It makes me sad that someone died there and it makes me want to be more careful	I always wonder who puts the flowers on them and how much pain they must be in
	I think they're nice especially since another person has taken the time to do it

DISCUSSION: RETURNING TO RECURSIVITY AND ROADSIDE CROSSES

Returning once more to the notion of recursivity, let me summarize how it illuminates the dynamics of this commemorative object and act. On the one hand, tangible and transient objects are embedded into the environment as individuals mark the site of an accident or where someone they loved died. With the establishment of the roadside cross and shrine, the environment is transformed or sanctified into a place of remembrance for the dead. What was once an ordinary place, unrecognized and undifferentiated, becomes recognized as a place of loss and grief. A cross or other marker placed at a particular location refers to itself and creates the location as a site of mourning or warning. Roadside crosses or other objects point (index) the relationship to what happened in the past at the current location. In a sense, the cross loops back and illuminates what happened at that particular site. It designates the site as a memorial and it consecrates (or sanctifies) the location as well. As such, there is an expansion of the field of meaning in the location and an increase in the level of complexity in the environment: the past is inserted into the present and gestures towards the future (i.e., "drive safely").

Furthermore, the deceased is also tangibly brought back into view, and as Santino (2006, p. 13) describes, the living "display death in the heart of social life." Functioning as a commentary or a message to the living, the cross and shrine maintain the deceased in the consciousness of community of the living.

More than simply recursively transforming the person/environment relationship or relating through the process of looping or embedding, crosses and shrines establish a connection of different levels of relationships and possibilities of relating. Memorial objects are evocative because they gather, enjoin, and connect the living and the dead in powerful ways. As boundary objects, they create and evoke multiple relationships and positions that make possible the unfolding and expansion of the imagination while the person is situated in the immediate field of experience.

In terms of the actor–network theory (ANT) (Latour, 2007), roadside memorials and crosses assemble together groups and collectivities. ANT seeks to understand how groups are connected, linked, and joined by social and nonsocial actors. Networks are complex groups comprised of heterogeneous elements that can be people, organizations, objects, or technologies. Thus, people and material objects are inseparable and therefore constitute each other. Importantly, ANT theorists do not see social groups as an ontological given and simply reproduced but rather something tenuous that is created and perpetuated through a myriad of practices, people, and objects. Indeed, people, organizations, and objects/artifacts are *actors* and can be said to have a degree of agency. Taken together with Herbst's co-genetic logic, one implication of these assertions is that the relationship between living and dead is fragile (especially in our modern times—see Westgaard, 2006), and the roadside memorial and cross do more than make grief manifest; rather, it is one possible means of creating and maintaining the relationship between the living and deceased and allow the living to relate to the absence of their departed loved ones.

CONCLUSION: A SIMPLE ACT, A COMPLICATED EFFECT

With the introduction of the interstate highway system, the landscape of the United States became more homogeneous. Roads intersecting towns and running through Main Street were replaced by miles of concrete that would make life more efficient. As the case with any road and its liminal nature, the interstate did not make travel safer, and signs of death and that commemorate tragic loss have flourished in recent times. The simple act of adding a cross or other form of memorial does more than simply add another object to the road; rather, it produces a nonlinear, recursive effect. One implication of this chapter involves the significance of material cultural objects; they are exclusively props or used by individuals to support their own development and to make sense of their world and to remember. Artifacts like memorial objects accomplish these goals and are interwoven with human ontogenesis. However, objects like roadside crosses create spaces through which we can reflect on abstract notions like death and move between different perspectives of those who are in mourning, those of are witnesses to mourning, and the dead. For most passing by, the roadside cross and shrine mark the death of someone unknown (a general other) but easily imagined—a young person killed while texting, a young mother killed by a drunk driver, a father killed on a dangerous road. Of course they hint that this might be our fate, but they also suggest and illuminate the relationship between the family and friends who left the marker and the deceased they wish to remember (and hope that *we* remember). This relationship is presented to the public. While not necessarily the primary focus, those who leave behind these markers are aware they will be viewed by many others and that their responses—someone slowing down, shedding a brief tear and remembering/knowing that someone died there—signify that their loved one is not completely lost because they exist in the mind of another, if only fleetingly. Roadside crosses and shrines accomplish much—they allow us to simultaneously be distanced from and united to the here-and-now setting via the imagination. At the heart of this phenomenon lies the recursive process.

REFERENCES

Aries, P. (1974). *Western attitudes toward death: From the middle ages to the present.* Baltimore, MD: Johns Hopkins University Press.

Beckstead, Z. (2012). Values internalization on the move: The revivification of faith along the pilgrims' path. In A. U. Branco & J. Valsiner (Eds.), *Cultural psychology of human values* (pp. 87–111). Charlotte, NC: Information Age Publishing.

Beckstead, Z., Twose, G., Levesque-Gottlieb, E., & Rizzo, J. (2011). Collective remembering through the materiality and organization of war memorials. *Journal of Material Culture 16*(2), 193–213.

Chomsky, N. (1957). *Syntactic structures.* The Hague, The Netherlands: Mouton.

Clifford, P. A., Friesen, S., & Jardine, D. (2001, April). *The ontology of hope: Lessons from a child.* Paper presented at the 82nd Annual Meeting of the American Educational Research Association, Seattle, WA. Retrieved from http://jan.ucc.nau.edu/ ~chaplx-p/docs/CliffordetalAERA2001Paper.pdf

Everett, H. (2002). *Roadside crosses in contemporary memorial culture.* Denton, TX: University of North Texas Press.

Foote, K. E. (2007). *Shadowed ground: America's landscape of violence and tragedy.* Austin, TX: The University of Texas Press.

Gillespie, A. (2006). *Becoming other: From social interaction to self-reflection.* Greenwich, CT: Information Age Publishing.

Herbst, D. (1995). What happens when we make a distinction: An elementary introduction to co-genetic logic. In T. Kindermann & J. Valsiner (Eds.), *Development of person-context relations* (pp. 67–80). Hillsdale, NJ: Lawrence Erlbaum Associates.

Latour, B. (2007). *Reassembling the social: An introduction to actor-network theory.* New York, NY: Oxford University Press.

Neuman, Y. (2003). *Processes and boundaries of the mind: Extending the limit.* Dordrecht, NL: Kluwer Academic/Plenum Publishers.

Owens, M. (2006). Louisiana roadside memorials: Negotiating an emerging tradition. In J. Santino (Ed.), *Spontaneous shrines and the public memorialization of death* (pp. 119–145). Houndsmills, Basingstroke, UK: Palmgrove Macmillan.

Pinker, S., & Jackendoff, R. (2005). The faculty of language: What's special about it? *Cognition,* 95, 201–236.

Santino, J. (2006). Performative commemoratives: Spontaneous shrines and the public memorialization of death. In J. Santino (Ed.), *Spontaneous Shrines and the Public Memorialization of Death* (pp. 5–15). Houndsmills, Basingstroke, UK: Palmgrove Macmillan.

Turner, V., & Turner, E. (1978). Image and pilgrimage in Christian culture. New York, NY: Columbia University Press.

Valsiner, J. (2007). *Culture in minds and societies: Foundations of cultural psychology.* Thousand Oaks, CA: Sage.

Wagner-Pacifici, R., & Scwartz, B. (1991). The Vietnam veterans memorial: Commemorating a difficult past. *American Journal of Sociology 2*(97), 376–420.

Weber, M. (1993). *The sociology of religion.* (E. Fischoff, Trans.). Boston, MA: Beacon Press. (Original work published in 1922)

Westgaard, H. (2006). "Like a trace": The spontaneous shrine as a cultural expression of grief. In J. Santino (Ed.), *Spontaneous shrines and the public memorialization of death* (pp. 147–175). Houndsmills, Basingstroke, UK: Palmgrove Macmillan.

Winter, J. (2006). *Remembering war: The great war between historical memory and history in the twentieth century*. New Haven, CT: Yale University Press.

GENERAL CONCLUSION

THE END IS THE BEGINNING

Moving Forward While Looking Backward

In true recursive fashion, we return to the beginning in order to review the discussions regarding recursivity. As Long et al mention in their chapter, given how central recursion and recursivity are to the human and social sciences, it is odd how infrequent these notions are explicitly used in their disciplines. Importantly, I believe, these chapters have successfully brought notions of recursivity and recursion into some unchartered areas of the social and human sciences. However, we have to acknowledge Rudolph's survey of recursion and recursivity where he demonstrates that when used in the social and human sciences, these ideas have multiple meanings, are confused, and are rarely clarified. Indeed, often the meaning of recursion in articles is left implicit without further clarification or development. One question that is mainly up to the reader is whether or not this volume presents a more systematic exploration of recursion and recursivity. While there is no single definition of recursion and recursivity universally adopted by all authors, I believe each has carefully articulated how they view and use recursivity and have contributed to both theoretical elaborations of recusrivity, and many have explored these notions in particular empirical contexts.

Still many questions remain: Do we simply "apply" notions of recursivity to new phenomena? Which models of recursivity are most useful in the

Cultural Psychology of Recursive Processes, pages 241–243

human and social sciences? Can we create new understandings of recursivity that are able to model dynamic, open systems? More importantly, do our explorations lead to further questions and generate new leads in studying phenomena and understanding recursivity? I believe this volume is a first step in addressing these complex questions.

By way of conclusion, I would like to make two points about recursivity and recursion that unite many of the chapters. First, I want to return to the first chapter in order to highlight an episode between researcher and participant that captures what I see as one of the most interesting and provocative aspects of recursion that run throughout the chapters in this book. The episode occurred as a mother and child were playing "eating soup" with real spoons while being videotaped. As the parent feeds the child after telling him that it was hot, the child eats and then looks at the researcher on the other side of the camera, giving a complicit smile. As Perinat notes, this is a creative act that draws in the researcher (observer) as a participant in the situation, and the participant becomes observer: The child creates the observer (researcher) by gazing at him or her and the child becomes the observer by noticing he is being observed.

> This is a self-referential act: Only an observer can see the other who is outside as an observer. In the children's first smiles of complicity directed towards those who are witnessing their shows or "as if" play, we sense the awakening of their observer's mind. However—and here is the superb paradox—the (human) observer is someone who, as we have seen above, is involved in the situation they observe. *The observer is "dragged" into the situation by the complicit gaze of the children. This is something that can only happen if there is an "expansion" of the scene,* which from this point on includes the new character. (Perinat, this volume, p. 4, emphasis added)

It is the highlighted sentences that I want to focus on: Here the recursive act *expands* the scene through insertion of the observer into the field of action and with the creation of new positions and perspectives for the researcher (now participant) and the child (both now observer and observed). The recursive act triggered a change, and more importantly, an *expansion* in the social relationships and the subjective worlds of the actors, and new possible courses of action and feeling are created. For instance, the if the mother catches the complicit smile of the child directed at the researcher, she might experience surprise at her child's growing cognitive capacities or embarrassed to be being "out the loop." In other words, recursive processes and actions open up the horizon of possibilities in both the *intra* and *inter* psychological realms. Like mirrors facing each other creating a vast series of images, our psyche has the potential for reaching toward time long since past, vistas never encountered, and minds seemingly inaccessible.

The second has to do with the relationship between recursion and imagination. While there are diverse threads and themes running throughout this book, one of the most explicit is the role of the human imagination, its expansive nature, and its recursive underpinnings[1]. It is human beings that find horror (or delight) being caught between two mirrors facing each other or contemplating bottomless turtles and infinity (Rudolph, Chapter 2). We bring voices of those deceased into our present-day conversations; we address and respond to them in creative ways that illuminate current problems (Charles, Chapter 3). Human beings nimbly move between the past, present, and future sometimes when encountering objects in our material environments (Beckstead, Chapter 8) or through the process of identity exploration (Watzlawik et al., Chapter 6). We anticipate how we will be evaluated by others and reveal and conceal our beliefs and identities accordingly (Long, Chapter 4). Language plays a key role in all of these recursive acts (Perinat, Chapter 1), yet we are deeply (i.e., materially or biologically) connected to others and thus have some sort of access to other minds (Guimaraes & Cravo, Chapter 5). Indeed, to paraphrase a well-known phrase by Clifford Geertz, human beings are suspended in webs of significance that they themselves recursively weave, and we do so in our various social roles (Cortes, Chapter 7): as psychologists, researchers, participants in the research process, children, parents, friends, and so on.

NOTE

1. We also must acknowledge that not all forms of repetition or looping (what we may call recursive acts or processes) are productive or adaptive. Long et al. argue that the recursive process is at play with "nones" who do not reveal their nonreligious identities but claim they would if asked (i.e., "hiding without hiding"). Furthermore, to borrow from the radical empiricist approach that Charles insightfully explores, not every act of return *is experienced* as novel or productive. Here we can cite Pierre Janet and his analysis of catastrophic thinking and ruminations in individuals (see Van Der Hart, Brown, & Van Der Kolk, 1989; Valsiner, 2007).

REFERENCES

Valsiner, J. (2007). *Culture in minds and societies.* Thousand Oaks, CA: Sage.

Van Der Hart, O., Brown, P., & Van Der Kolk, B. A. (1989). Pierre Janet's treatment of post-traumatic stress. *Journal of Traumatic Stress* 2(4), 379–395.

Printed and bound by CPI Group (UK) Ltd, Croydon, CR0 4YY

15/09/2026

14962338-0004